High Voltage Engineering

High Voltage Engineering

Edited by
Peter Mackintosh

Larsen & Keller
www.larsen-keller.com

High Voltage Engineering
Edited by Peter Mackintosh
ISBN: 978-1-63549-689-5 (Hardback)

© 2018 Larsen & Keller

Larsen & Keller

Published by Larsen and Keller Education,
5 Penn Plaza,
19th Floor,
New York, NY 10001, USA

Cataloging-in-Publication Data

High voltage engineering / edited by Peter Mackintosh.
    p. cm.
Includes bibliographical references and index.
ISBN 978-1-63549-689-5
    1. Electric power distribution--High tension. 2. High voltages. 3. Electrical engineering.
I. Mackintosh, Peter.
TK3144 .H54 2018
621.319 13--dc23

For more information regarding Larsen and Keller Education and its products, please visit the publisher's website www.larsen-keller.com

Table of Contents

Preface

As a sub-field of electrical engineering, high voltage engineering refers to the study and understanding of the different levels of voltage and how they affect the technology and especially, electricity transmission. The various levels of voltage include low, medium, high, extra and ultra high voltage. The subject also includes high voltage generation, its consumption, transmission, distribution and also, its applications in the industrial sector. The book studies, analyses and uphold the pillars of high voltage engineering and its utmost significance in modern times. It is a valuable compilation of topics, ranging from the basic to the most complex theories and principles in this field. For all those who are interested in high voltage engineering, this textbook can prove to be an essential guide.

To facilitate a deeper understanding of the contents of this book a short introduction of every chapter is written below:

Chapter 1- High voltage causes immense damage to life and property. It is used in distributing power to electrical appliances, generating x-rays and in photomultiplier tubes. The chapter on high voltage offers an insightful focus, keeping in mind the complex subject matter.

Chapter 2- Dielectrics are material that can store electrostatic charge. The principles of dielectrics are fundamental in the development of the capacitor. Dielectrics are an emerging field of study; the following section will not only provide an overview, it will also delve deep into the variegated topics related to it.

Chapter 3- Solid dielectric materials such as porcelain, mica, plastics, etc. are used when current is transferred within electrical circuits containing different voltages. Liquid dielectrics on the other hand are used to prevent electrical discharge. The topics discussed in the chapter are of great importance to broaden the existing knowledge on liquid and solid dielectrics.

Chapter 4- In order to measure different types of voltages there are laboratory techniques. The different types of voltages are ac power frequency, dc voltages and impulse voltages. The chapter closely examines high voltage measurement to provide an extensive understanding of the subject.

Chapter 5- A transformer conducts electricity from one place to another through electromagnetic induction. It allows electrical current to be adjusted and regulated. The aspects elucidated in this chapter are of vital importance, and provide a better understanding of high voltage engineering.

I would like to share the credit of this book with my editorial team who worked tirelessly on this book. I owe the completion of this book to the never-ending support of my family, who supported me throughout the project.

Editor

Understanding High Voltage

High voltage causes immense damage to life and property. It is used in distributing power to electrical appliances, generating x-rays and in photomultiplier tubes. The chapter on high voltage offers an insightful focus, keeping in mind the complex subject matter.

High Voltage

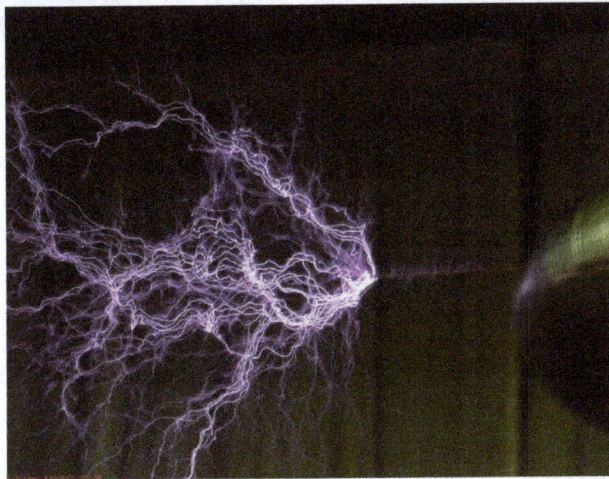

High voltages may lead to electrical breakdown, resulting in an electrical discharge as illustrated by the plasma filaments streaming from a Tesla coil.

The term high voltage usually means electrical energy at voltages high enough to inflict harm on living organisms. Equipment and conductors that carry high voltage warrant particular safety requirements and procedures. In certain industries, *high voltage* means voltage above a particular threshold. High voltage is used in electrical power distribution, in cathode ray tubes, to generate X-rays and particle beams, to demonstrate arcing, for ignition, in photomultiplier tubes, and in high power amplifier vacuum tubes and other industrial and scientific applications.

Definition

IEC Voltage Range	AC (V_{rms})	DC (V)	Defining Risk
High voltage (supply system)	> 1000	> 1500	Electrical arcing
Low voltage (supply system)	50–1000	120–1500	Electrical shock
Extra-low voltage (supply system)	< 50	< 120	Low risk

The numerical definition of 'high voltage' depends on context. Two factors considered in classifying a voltage as "high voltage" are the possibility of causing a spark in air, and the danger of electric shock by contact or proximity. The definitions may refer to the voltage between two conductors of a system, or between any conductor and ground.

In electric power transmission engineering, high voltage is usually considered any voltage over approximately 35,000 volts. This is a classification based on the design of apparatus and insulation.

The International Electrotechnical Commission and its national counterparts (IET, IEEE, VDE, etc.) define *high voltage* as above 1000 V for alternating current, and at least 1500 V for direct current—and distinguish it from low voltage (50–1000 V AC or 120–1500 V DC) and extra-low voltage (<50 V AC or <120 V DC) circuits. This is in the context of building wiring and the safety of electrical apparatus.

In the United States 2011 National Electrical Code (NEC) is the standard regulating most electrical installations. There are no definitions relating to *high voltage*. The NEC® covers voltages 600 volts and less and that over 600 volts. The National Electrical Manufacturer's Association (NEMA) defines high voltage as over 100kV to 230kV. British Standard BS 7671:2008 defines *high voltage* as any voltage difference between conductors that is higher than 1000 V AC or 1500 V ripple-free DC, or any voltage difference between a conductor and Earth that is higher than 600 V AC or 900 V ripple-free DC.

Electricians may only be licensed for particular voltage classes, in some jurisdictions. For example, an electrical license for a specialized sub-trade such as installation of HVAC systems, fire alarm systems, closed circuit television systems may be authorized to install systems energized up to only 30 volts between conductors, and may not be permitted to work on mains-voltage circuits. The general public may consider household mains circuits (100–250 V AC), which carry the highest voltages they normally encounter, to be *high voltage.*

Voltages over approximately 50 volts can usually cause dangerous amounts of current to flow through a human being who touches two points of a circuit—so safety standards, in general, are more restrictive around such circuits.. The definition of *extra high voltage* (EHV) again depends on context. In electric power transmission engineering, EHV equipment carries more than 345,000 volts between conductors. In electronics systems, a power supply that provides greater than 275,000 volts is called an *EHV Power Supply*, and is often used in experiments in physics.

The accelerating voltage for a television cathode ray tube may be described as *extra-high voltage* or *extra-high tension* (EHT), compared to other voltage supplies within the equipment. This type of supply ranges from >5 kV to about 50 kV.

In automotive engineering, high voltage is defined as voltage in range 30–1000 V_{ac} or 60–1500 V_{dc}.

In digital electronics, a high voltage usually refers to that representing a logic 1 in positive logic and a logic 0 in negative logic. It is not used to indicate a hazardous voltage and levels between ICs to TTL/CMOS standards and their modern derivatives are well below hazardous levels. The highest in mainstream use were 15V for original CMOS and 5V for TTL but modern devices use 3.3V, with 1.8V or lower used in many applications.

Safety

Voltages greater than 50 V applied across dry unbroken human skin can cause heart fibrillation if they produce electric currents in body tissues that happen to pass through the chest area. The voltage at which there is the danger of electrocution depends on the electrical conductivity of dry human skin. Living human tissue can be protected from damage by the insulating characteristics of dry skin up to around 50 volts. If the same skin becomes wet, if there are wounds, or if the voltage is applied to electrodes that penetrate the skin, then even voltage sources below 40 V can be lethal.

International safety symbol "Caution, risk of electric shock" (ISO 3864), also known as *high voltage symbol*.

Accidental contact with high voltage supplying sufficient energy may result in severe injury or death. This can occur as a person's body provides a path for current flow, causing tissue damage and heart failure. Other injuries can include burns from the arc generated by the accidental contact. These burns can be especially dangerous if the victim's airways are affected. Injuries may also be suffered as a result of the physical forces experienced by people who fall from a great height or are thrown a considerable distance.

Low-energy exposure to high voltage may be harmless, such as the spark produced in a dry climate when touching a doorknob after walking across a carpeted floor. The voltage can be in the thousand-volt range, but the current (the rate of charge transfer) is low.

Safety equipment used by electrical workers includes insulated rubber gloves and mats. These protect the user from electric shock. Safety equipment is tested regularly to ensure it is still protecting the user. Test regulations vary according to country. Testing companies can test at up 300,000 volts and offer services from glove testing to Elevated Working Platform (or EWP) testing.

Sparks in Air

The dielectric breakdown strength of dry air, at Standard Temperature and Pressure (STP), between spherical electrodes is approximately 33 kV/cm. This is only as a rough guide, since the actual breakdown voltage is highly dependent upon the electrode shape and size. Strong electric fields (from high voltages applied to small or pointed conductors) often produce violet-colored corona discharges in air, as well as visible sparks. Voltages below about 500–700 volts cannot produce easily visible sparks or glows in air at atmospheric pressure, so by this rule these voltages are "low". However, under conditions of low atmospheric pressure (such as in high-altitude aircraft), or in an environment of noble gas such as argon or neon, sparks appear at much lower voltages. 500 to 700 volts is not a fixed minimum for producing spark breakdown, but it is a rule-of-thumb. For air at STP, the minimum sparkover voltage is around 327 volts, as noted by Friedrich Paschen.

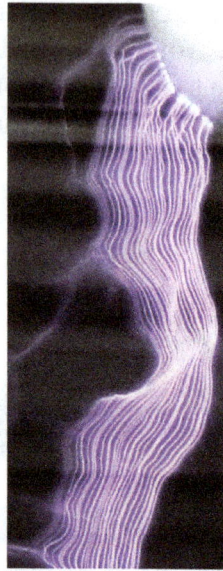

Long exposure photograph of a Tesla coil showing the repeated electric discharges

While lower voltages do not, in general, jump a gap that is present before the voltage is applied, interrupting an existing current flow with a gap often produces a low-voltage spark or arc. As the contacts are separated, a few small points of contact become the last to separate. The current becomes constricted to these small *hot spots*, causing them to become incandescent, so that they emit electrons (through thermionic emission). Even a small 9 V battery can spark noticeably by this mechanism in a darkened room. The ionized air and metal vapour (from the contacts) form plasma, which temporarily bridges the widening gap. If the power supply and load allow sufficient current to flow, a self-sustaining arc may form. Once formed, an arc may be extended to a significant length before breaking the circuit. Attempting to open an inductive circuit often forms an arc, since the inductance provides a high-voltage pulse whenever the current is interrupted. AC systems make sustained arcing somewhat less likely, since the current returns to zero twice per cycle. The arc is extinguished every time the current goes through a zero crossing, and must reignite during the next half-cycle to maintain the arc.

Unlike an ohmic conductor, the resistance of an arc decreases as the current increases. This makes unintentional arcs in an electrical apparatus dangerous since even a small arc can grow large enough to damage equipment and start fires if sufficient current is available. Intentionally produced arcs, such as used in lighting or welding, require some element in the circuit to stabilize the arc's current/voltage characteristics.

Electrostatic Devices, Natural Static Electricity and Similar Phenomena

A high voltage is not necessarily dangerous if it cannot deliver substantial current. The common static electric sparks seen under low-humidity conditions always involve voltage well above 700 V. For example, sparks to car doors in winter can involve voltages as high as 20,000 V. Also, physics demonstration devices such as Van de Graaff generators and Wimshurst machines can produce voltages approaching one million volts, yet at worst they deliver a brief sting. That is because the number of electrons involved is not high. These devices have a limited amount of stored energy, so the average current produced is low and usually for a short time, with impulses peaking in the amp range for a nanosecond. During the discharge, these machines apply high voltage to the body for only a millionth of a second or less. So a low-amperage current is applied for a very short time, and the number of electrons involved is very small.

The discharge may involve extremely high voltage over very short periods, but, to produce heart fibrillation, an electric power supply must produce a significant current (amperage) in the heart muscle continuing for many milliseconds, and must deposit a total energy in the range of at least millijoules or higher. A current of relatively high amperage at anything more than about fifty volts can therefore be medically significant and potentially fatal.

Tesla coils are not electrostatic machines and can produce significant currents for a sustained interval. Although their appearance in operation is similar to high voltage static electricity devices, the current supplied to a human body will be relatively constant as long as contact is maintained, and the voltage will be much higher than the break-down voltage of human skin. As a consequence, the output of a Tesla coil can be dangerous or even fatal.

Power Lines

Power lines with high voltage warning sign

Electrical transmission and distribution lines for electric power always use voltages significantly higher than 50 volts, so contact with or close approach to the line conductors presents a danger of electrocution. Contact with overhead wires is a frequent cause of injury or death. Metal ladders, farm equipment, boat masts, construction machinery, aerial antennas, and similar objects are frequently involved in fatal contact with overhead wires. Digging into a buried cable can also be dangerous to workers at an excavation site. Digging equipment (either hand tools or machine driven) that contacts a buried cable may energize piping or the ground in the area, resulting in electrocution of nearby workers. A fault in a high-voltage transmission line or substation may result in high currents flowing along the surface of the earth, producing an earth potential rise that also presents a danger of electric shock.

Unauthorized persons climbing on power pylons or electrical apparatus are also frequently the victims of electrocution. At very high transmission voltages even a close approach can be hazardous, since the high voltage may arc across a significant air gap.

For high-voltage and extra-high-voltage transmission lines, specially trained personnel use "live line" techniques to allow hands-on contact with energized equipment. In this case the worker is electrically connected to the high-voltage line but thoroughly insulated from the earth so that he is at the same electrical potential as that of the line. Since training for such operations is lengthy, and still presents a danger to personnel, only very important transmission lines are subject to maintenance while live. Outside these properly engineered situations, insulation from earth does not guarantee that no current flows to earth—as grounding or arcing to ground can occur in unexpected ways, and high-frequency currents can burn even an ungrounded person. Touching a transmitting antenna is dangerous for this reason, and a high-frequency Tesla coil can sustain a spark with only one endpoint.

Protective equipment on high-voltage transmission lines normally prevents formation of an unwanted arc, or ensures that it is quenched within tens of milliseconds. Electrical apparatus that interrupts high-voltage circuits is designed to safely direct the resulting arc so that it dissipates without damage. High voltage circuit breakers often use a blast of high pressure air, a special dielectric gas (such as SF_6 under pressure), or immersion in mineral oil to quench the arc when the high voltage circuit is broken.

Arc Flash Hazard

High voltage testing arrangement with large capacitor and test transformer

Depending on the prospective short circuit current available at a switchgear line-up, a hazard is presented to maintenance and operating personnel due to the possibility of a high-intensity electric arc. Maximum temperature of an arc can exceed 10,000 kelvin, and the radiant heat, expanding hot air, and explosive vaporization of metal and insulation material can cause severe injury to unprotected workers. Such switchgear line-ups and high-energy arc sources are commonly present in electric power utility substations and generating stations, industrial plants and large commercial buildings. In the United States, the National Fire Protection Association, has published a guideline standard NFPA 70E for evaluating and calculating *arc flash hazard*, and provides standards for the protective clothing required for electrical workers exposed to such hazards in the workplace.

Explosion Hazard

Even voltages insufficient to break down air can be associated with enough energy to ignite atmospheres containing flammable gases or vapours, or suspended dust. For example, hydrogen gas, natural gas, or petrol/gasoline vapor mixed with air can be ignited by sparks produced by electrical apparatus. Examples of industrial facilities with hazardous areas are petrochemical refineries, chemical plants, grain elevators, and coal mines.

Measures taken to prevent such explosions include:

- Intrinsic safety by the use of apparatus designed not to accumulate enough stored electrical energy to trigger an explosion.

- Increased safety, which applies to devices using measures such as oil-filled enclosures to prevent sparks.

- Explosion-proof (flame-proof) enclosures, which are designed so that an explosion within the enclosure cannot escape and ignite a surrounding explosive atmosphere (this designation does not imply that the apparatus can survive an internal or external explosion).

In recent years, standards for explosion hazard protection have become more uniform between European and North American practice. The "zone" system of classification is now used in modified form in U.S. National Electrical Code and in the Canadian Electrical Code. Intrinsic safety apparatus is now approved for use in North American applications.

Toxic Gases

Electrical discharges, including partial discharge and corona, can produce small quantities of toxic gases, which in a confined space can be a serious health hazard. These gases include ozone and various oxides of nitrogen.

Lightning

The largest scale sparks are those produced naturally by lightning. An average bolt of negative lightning carries a current of 30 to 50 kiloamperes, transfers a charge of 5 coulombs, and dissipates 500 megajoules of energy (120 kg TNT equivalent, or enough to light a 100-watt light bulb for approximately 2 months). However, an average bolt of positive lightning (from the top of a thunderstorm) may carry a current of 300 to 500 kiloamperes, transfer a charge of up to 300 coulombs, have a potential difference up to 1 gigavolt (a billion volts), and may dissipate 300 GJ of energy (72 tons TNT, or enough energy to light a 100-watt light bulb for up to 95 years). A negative lightning strike typically lasts for only tens of microseconds, but multiple strikes are common. A positive lightning stroke is typically a single event. However, the larger peak current may flow for hundreds of milliseconds, making it considerably hotter and more dangerous than negative lightning.

Hazards due to lightning obviously include a direct strike on persons or property. However, lightning can also create dangerous voltage gradients in the earth, as well as an electromagnetic pulse, and can charge extended metal objects such as telephone cables, fences, and pipelines to dangerous voltages that can be carried many miles from the site of the strike. Although many of these objects are not normally conductive, very high voltage can cause the electrical breakdown of such insulators, causing them to act as conductors. These transferred potentials are dangerous to people, livestock, and electronic apparatus. Lightning strikes also start fires and explosions, which result in fatalities, injuries, and property damage. For example, each year in North America, thousands of forest fires are started by lightning strikes.

Measures to control lightning can mitigate the hazard; these include lightning rods, shielding wires, and bonding of electrical and structural parts of buildings to form a continuous enclosure.

High-voltage lightning discharges in the atmosphere of Jupiter are thought to be the source of the planet's powerful radio frequency emissions.

Levels of High Voltage

World over the levels are classified as:

- LOW

- MEDIUM

- HIGH

- EXTRA and

- ULTRA HIGH Voltages

- However, the exact magnitude of these levels vary from country to country. Hence this system of technical terms for the voltage levels is inappropriate.

- In most part of the world even 440 V is considered to be high voltage since it is dangerous for the living being.

- Hence it would be more appropriate to always mention the level of voltagebeing referred without any set nomenclature.

Voltage Levels

Consumer

- ac power frequency :

  110 V, 220 V- single phase

  440 V, 3.3 kV ,6.6 kV, 11 kV-three phase (3.3 & 6.6 kV are being phased out).

- Besides these levels ,the Railway Traction at 25 kV , single phase is one of the biggest consumer of power spread at any particular stretch to 40 km of track length.

Generation :

  Three phase synchronous generators

  440 V, 3.3 kV, 6.6 kV (small generators) , 11 kV (110 & 220 MW).

21.5 kV (500 MW), 33 kV (1000 M

[limitation due to machine insulation requirement].

Distribution :

Three phase

440 V, 3.3 kV, 6.6 kV, 11 kV, 33 kV, 66 kV

With the increase in power consumption density, the power distribution voltage levels are at rise because the power handling capacity is proportional to the square of the voltage level.

(In Germany 440 V , 3.0 kV 6.0 kV, 10 kV, 30 kV, 60 kV).

AC Transmission :

110 kV, 132 kV, 220 kV, 380 - 400 kV, 500 kV, 765 - 800 kV, 1000 kV and 1150 kV exist. Work on 1500 kV is complete.

In three phase power system, the rated voltage is always given as line to line, rms voltage.

DC Transmission : dc single pole and bipolar lines :

± 100 kV to ± 500 kV

Advance countries like US, Canada and Japan have their single phase ac power consumption level at 110 V. Rest of the whole world consumes single phase ac power at 220 V.

The only advantage of 110 V single phase consumer voltage is that it is safer over 220 V. However, the disadvantages are many.

Disadvantages :

- It requires double the magnitude of current to deliver the same amount of power as at 220 V.

- Hence for the same magnitude of I2R losses to limit the conductor or the insulation temperature to 70° C (for PVC) , the resistance of the distribution cable should be 4 times lower. Therefore, the cable cross-section area has to be increased four folds.

- Four times more copper requirement, dumped in the building walls is an expensive venture.

- Due to higher magnitude of current, higher magnetic field in the buildings. Not good for health.

- With the installation of modern inexpensive protective devices (earth fault relays), 220 V is equally safe as 110 V.

Rated maximum temperature of cables:

- It is important to understand the current and voltage carrying capacities of a conductor separately. While the current carrying capability is determined by the conductivity of the conductors, directly proportional to the area of conductor cross-section, the voltage bearing capacity depends upon the level of insulation provided to the conductor.

- The current carrying capability in turn is determined by maximum permissible temperature of the insulation or that of the conductor.

- The real power loss, I^2R and the rate of cooling determine the temperature rise of the conductor which should not be more than the maximum permissible temperature of the type of insulation provided on the conductor.

- Hence, not only electrical but thermal and mechanical properties of insulation are important in power system.

Electricity

Lightning is one of the most dramatic effects of electricity

Electricity is the set of physical phenomena associated with the presence of electric charge. Although initially considered a phenomenon separate from magnetism, since the development of Maxwell's Equations, both are recognized as part of a single phenomenon: electromagnetism. Various common phenomena are related to electricity, including lightning, static electricity, electric heating, electric discharges and many others. In addition, electricity is at the heart of many modern technologies.

The presence of an electric charge, which can be either positive or negative, produces an electric field. On the other hand, the movement of electric charges, which is known as electric current, produces a magnetic field.

When a charge is placed in a location with non-zero electric field, a force will act on it. The magnitude of this force is given by Coulomb's Law. Thus, if that charge were to move, the electric field would be doing work on the electric charge. Thus we can speak of electric potential at a certain point in space, which is equal to the work done by an external agent in carrying a unit of positive charge from an arbitrarily chosen reference point to that point without any acceleration and is typically measured in volts.

In electrical engineering, electricity is used for:

- electric power where electric current is used to energise equipment;

- electronics which deals with electrical circuits that involve active electrical components such as vacuum tubes, transistors, diodes and integrated circuits, and associated passive interconnection technologies.

Electrical phenomena have been studied since antiquity, though progress in theoretical understanding remained slow until the seventeenth and eighteenth centuries. Even then, practical applications for electricity were few, and it would not be until the late nineteenth century that engineers were able to put it to industrial and residential use. The rapid expansion in electrical technology at this time transformed industry and society. Electricity's extraordinary versatility means it can be put to an almost limitless set of applications which include transport, heating, lighting, communications, and computation. Electrical power is now the backbone of modern industrial society.

History

Thales, the earliest known researcher into electricity

Long before any knowledge of electricity existed, people were aware of shocks from electric fish. Ancient Egyptian texts dating from 2750 BCE referred to these fish as the "Thunderer of the Nile", and described them as the "protectors" of all other fish. Electric fish were again reported millennia later by ancient Greek, Roman and Arabic naturalists and physicians. Several ancient writers, such as Pliny the Elder and Scribonius Largus, attested to the numbing effect of electric shocks delivered by catfish and electric rays, and knew that such shocks could travel along conducting objects. Patients suffering from ailments such as gout or headache were directed to touch electric fish in the hope that the powerful jolt might cure them. Possibly the earliest and nearest approach to the discovery of the identity of lightning, and electricity from any other source, is to be attributed to the Arabs, who before the 15th century had the Arabic word for lightning *ra'ad* applied to the electric ray.

Ancient cultures around the Mediterranean knew that certain objects, such as rods of amber, could be rubbed with cat's fur to attract light objects like feathers. Thales of Miletus made a series of observations on static electricity around 600 BCE, from which he believed that friction rendered amber magnetic, in contrast to minerals such as magnetite, which needed no rubbing. Thales was incorrect in believing the attraction was due to a magnetic effect, but later science would prove a link between magnetism and electricity. According to a controversial theory, the

Parthians may have had knowledge of electroplating, based on the 1936 discovery of the Baghdad Battery, which resembles a galvanic cell, though it is uncertain whether the artifact was electrical in nature.

Benjamin Franklin conducted extensive research on electricity in the 18th century, as documented by Joseph Priestley (1767) *History and Present Status of Electricity*, with whom Franklin carried on extended correspondence.

Electricity would remain little more than an intellectual curiosity for millennia until 1600, when the English scientist William Gilbert made a careful study of electricity and magnetism, distinguishing the lodestone effect from static electricity produced by rubbing amber. He coined the New Latin word *electricus* to refer to the property of attracting small objects after being rubbed. This association gave rise to the English words "electric" and "electricity", which made their first appearance in print in Thomas Browne's *Pseudodoxia Epidemica* of 1646.

Further work was conducted by Otto von Guericke, Robert Boyle, Stephen Gray and C. F. du Fay. In the 18th century, Benjamin Franklin conducted extensive research in electricity, selling his possessions to fund his work. In June 1752 he is reputed to have attached a metal key to the bottom of a dampened kite string and flown the kite in a storm-threatened sky. A succession of sparks jumping from the key to the back of his hand showed that lightning was indeed electrical in nature. He also explained the apparently paradoxical behavior of the Leyden jar as a device for storing large amounts of electrical charge in terms of electricity consisting of both positive and negative charges.

Michael Faraday's discoveries formed the foundation of electric motor technology

In 1791, Luigi Galvani published his discovery of bioelectromagnetics, demonstrating that electricity was the medium by which neurons passed signals to the muscles. Alessandro Volta's battery, or voltaic pile, of 1800, made from alternating layers of zinc and copper, provided scientists with a more reliable source of electrical energy than the electrostatic machines previously used. The recognition of electromagnetism, the unity of electric and magnetic phenomena, is due to Hans Christian Ørsted and André-Marie Ampère in 1819–1820. Michael Faraday invented the electric motor in 1821, and Georg Ohm mathematically analysed the electrical circuit in 1827. Electricity and magnetism (and light) were definitively linked by James Clerk Maxwell, in particular in his "On Physical Lines of Force" in 1861 and 1862.

While the early 19th century had seen rapid progress in electrical science, the late 19th century would see the greatest progress in electrical engineering. Through such people as Alexander Graham Bell, Ottó Bláthy, Thomas Edison, Galileo Ferraris, Oliver Heaviside, Ányos Jedlik, William Thomson, 1st Baron Kelvin, Charles Algernon Parsons, Werner von Siemens, Joseph Swan, Reginald Fessenden, Nikola Tesla and George Westinghouse, electricity turned from a scientific curiosity into an essential tool for modern life, becoming a driving force of the Second Industrial Revolution.

In 1887, Heinrich Hertz discovered that electrodes illuminated with ultraviolet light create electric sparks more easily. In 1905, Albert Einstein published a paper that explained experimental data from the photoelectric effect as being the result of light energy being carried in discrete quantized packets, energising electrons. This discovery led to the quantum revolution. Einstein was awarded the Nobel Prize in Physics in 1921 for "his discovery of the law of the photoelectric effect". The photoelectric effect is also employed in photocells such as can be found in solar panels and this is frequently used to make electricity commercially.

The first solid-state device was the "cat's-whisker detector" first used in the 1900s in radio receivers. A whisker-like wire is placed lightly in contact with a solid crystal (such as a germanium crystal) in order to detect a radio signal by the contact junction effect. In a solid-state component, the current is confined to solid elements and compounds engineered specifically to switch and amplify it. Current flow can be understood in two forms: as negatively charged electrons, and as positively charged electron deficiencies called holes. These charges and holes are understood in terms of quantum physics. The building material is most often a crystalline semiconductor.

The solid-state device came into its own with the invention of the transistor in 1947. Common solid-state devices include transistors, microprocessor chips, and RAM. A specialized type of RAM called flash RAM is used in USB flash drives and more recently, solid-state drives to replace mechanically rotating magnetic disc hard disk drives. Solid state devices became prevalent in the 1950s and the 1960s, during the transition from vacuum tubes to semiconductor diodes, transistors, integrated circuit (IC) and the light-emitting diode (LED).

Concepts

Electric Charge

The presence of charge gives rise to an electrostatic force: charges exert a force on each other, an effect that was known, though not understood, in antiquity. A lightweight ball suspended from a string can be charged by touching it with a glass rod that has itself been charged by rubbing with a cloth. If a similar ball is charged by the same glass rod, it is found to repel the first: the charge acts to force the two balls apart. Two balls that are charged with a rubbed amber rod also repel each other. However, if one ball is charged by the glass rod, and the other by an amber rod, the two balls are found to attract each other. These phenomena were investigated in the late eighteenth century by Charles-Augustin de Coulomb, who deduced that charge manifests itself in two opposing forms. This discovery led to the well-known axiom: *like-charged objects repel and opposite-charged objects attract.*

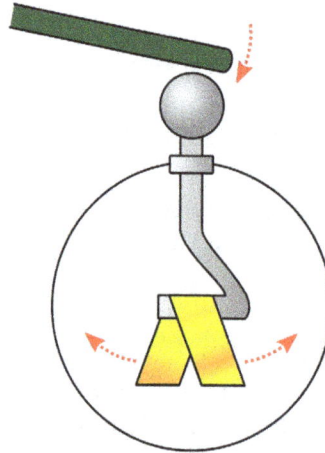

Charge on a gold-leaf electroscope causes the leaves to visibly repel each other

The force acts on the charged particles themselves, hence charge has a tendency to spread itself as evenly as possible over a conducting surface. The magnitude of the electromagnetic force, whether attractive or repulsive, is given by Coulomb's law, which relates the force to the product of the charges and has an inverse-square relation to the distance between them. The electromagnetic force is very strong, second only in strength to the strong interaction, but unlike that force it operates over all distances. In comparison with the much weaker gravitational force, the electromagnetic force pushing two electrons apart is 10^{42} times that of the gravitational attraction pulling them together.

Study has shown that the origin of charge is from certain types of subatomic particles which have the property of electric charge. Electric charge gives rise to and interacts with the electromagnetic force, one of the four fundamental forces of nature. The most familiar carriers of electrical charge are the electron and proton. Experiment has shown charge to be a conserved quantity, that is, the net charge within an isolated system will always remain constant regardless of any changes taking place within that system. Within the system, charge may be transferred between bodies, either by direct contact, or by passing along a conducting material, such as a wire. The informal term static electricity refers to the net presence (or 'imbalance') of charge on a body, usually caused when dissimilar materials are rubbed together, transferring charge from one to the other.

The charge on electrons and protons is opposite in sign, hence an amount of charge may be expressed as being either negative or positive. By convention, the charge carried by electrons is deemed negative, and that by protons positive, a custom that originated with the work of Benjamin Franklin. The amount of charge is usually given the symbol Q and expressed in coulombs; each electron carries the same charge of approximately -1.6022×10^{-19} coulomb. The proton has a charge that is equal and opposite, and thus $+1.6022\times10^{-19}$ coulomb. Charge is possessed not just by matter, but also by anti-matter, each antiparticle bearing an equal and opposite charge to its corresponding particle.

Charge can be measured by a number of means, an early instrument being the gold-leaf electroscope, which although still in use for classroom demonstrations, has been superseded by the electronic electrometer.

Electric Current

The movement of electric charge is known as an electric current, the intensity of which is usually measured in amperes. Current can consist of any moving charged particles; most commonly these are electrons, but any charge in motion constitutes a current. Electric current can flow through some things, electrical conductors, but will not flow through an electrical insulator.

An electric arc provides an energetic demonstration of electric current

By historical convention, a positive current is defined as having the same direction of flow as any positive charge it contains, or to flow from the most positive part of a circuit to the most negative part. Current defined in this manner is called conventional current. The motion of negatively charged electrons around an electric circuit, one of the most familiar forms of current, is thus deemed positive in the *opposite* direction to that of the electrons. However, depending on the conditions, an electric current can consist of a flow of charged particles in either direction, or even in both directions at once. The positive-to-negative convention is widely used to simplify this situation.

The process by which electric current passes through a material is termed electrical conduction, and its nature varies with that of the charged particles and the material through which they are travelling. Examples of electric currents include metallic conduction, where electrons flow through a conductor such as metal, and electrolysis, where ions (charged atoms) flow through liquids, or through plasmas such as electrical sparks. While the particles themselves can move quite slowly, sometimes with an average drift velocity only fractions of a millimetre per second, the electric field that drives them itself propagates at close to the speed of light, enabling electrical signals to pass rapidly along wires.

Current causes several observable effects, which historically were the means of recognising its presence. That water could be decomposed by the current from a voltaic pile was discovered by Nicholson and Carlisle in 1800, a process now known as electrolysis. Their work was greatly expanded upon by Michael Faraday in 1833. Current through a resistance causes localised heating, an effect James Prescott Joule studied mathematically in 1840. One of the most important discoveries relating to current was made accidentally by Hans Christian Ørsted in 1820, when, while preparing a lecture, he witnessed the current in a wire disturbing the needle of a magnetic compass. He had discovered electromagnetism, a fundamental interaction between electricity and magnetics. The level of electromagnetic emissions generated by electric arcing is high enough to produce electromagnetic interference, which can be detrimental to the workings of adjacent equipment.

In engineering or household applications, current is often described as being either direct current (DC) or alternating current (AC). These terms refer to how the current varies in time. Direct current, as produced by example from a battery and required by most electronic devices, is a unidirectional flow from the positive part of a circuit to the negative. If, as is most common, this flow is carried by electrons, they will be travelling in the opposite direction. Alternating current is any current that reverses direction repeatedly; almost always this takes the form of a sine wave. Alternating current thus pulses back and forth within a conductor without the charge moving any net distance over time. The time-averaged value of an alternating current is zero, but it delivers energy in first one direction, and then the reverse. Alternating current is affected by electrical properties that are not observed under steady state direct current, such as inductance and capacitance. These properties however can become important when circuitry is subjected to transients, such as when first energised.

Electric Field

The concept of the electric field was introduced by Michael Faraday. An electric field is created by a charged body in the space that surrounds it, and results in a force exerted on any other charges placed within the field. The electric field acts between two charges in a similar manner to the way that the gravitational field acts between two masses, and like it, extends towards infinity and shows an inverse square relationship with distance. However, there is an important difference. Gravity always acts in attraction, drawing two masses together, while the electric field can result in either attraction or repulsion. Since large bodies such as planets generally carry no net charge, the electric field at a distance is usually zero. Thus gravity is the dominant force at distance in the universe, despite being much weaker.

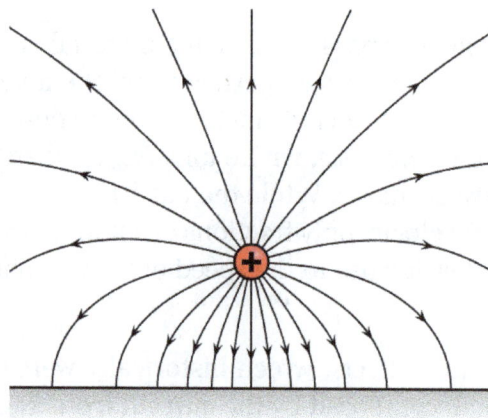

Field lines emanating from a positive charge above a plane conductor

An electric field generally varies in space, and its strength at any one point is defined as the force (per unit charge) that would be felt by a stationary, negligible charge if placed at that point. The conceptual charge, termed a 'test charge', must be vanishingly small to prevent its own electric field disturbing the main field and must also be stationary to prevent the effect of magnetic fields. As the electric field is defined in terms of force, and force is a vector, so it follows that an electric field is also a vector, having both magnitude and direction. Specifically, it is a vector field.

The study of electric fields created by stationary charges is called electrostatics. The field may be visualised by a set of imaginary lines whose direction at any point is the same as that of the field. This concept was introduced by Faraday, whose term 'lines of force' still sometimes sees use. The field lines are the paths that a point positive charge would seek to make as it was forced to move within the field; they are however an imaginary concept with no physical existence, and the field permeates all the intervening space between the lines. Field lines emanating from stationary charges have several key properties: first, that they originate at positive charges and terminate at negative charges; second, that they must enter any good conductor at right angles, and third, that they may never cross nor close in on themselves.

A hollow conducting body carries all its charge on its outer surface. The field is therefore zero at all places inside the body. This is the operating principal of the Faraday cage, a conducting metal shell which isolates its interior from outside electrical effects.

The principles of electrostatics are important when designing items of high-voltage equipment. There is a finite limit to the electric field strength that may be withstood by any medium. Beyond this point, electrical breakdown occurs and an electric arc causes flashover between the charged parts. Air, for example, tends to arc across small gaps at electric field strengths which exceed 30 kV per centimetre. Over larger gaps, its breakdown strength is weaker, perhaps 1 kV per centimetre. The most visible natural occurrence of this is lightning, caused when charge becomes separated in the clouds by rising columns of air, and raises the electric field in the air to greater than it can withstand. The voltage of a large lightning cloud may be as high as 100 MV and have discharge energies as great as 250 kWh.

The field strength is greatly affected by nearby conducting objects, and it is particularly intense when it is forced to curve around sharply pointed objects. This principle is exploited in the lightning conductor, the sharp spike of which acts to encourage the lightning stroke to develop there, rather than to the building it serves to protect.

Electric Potential

The concept of electric potential is closely linked to that of the electric field. A small charge placed within an electric field experiences a force, and to have brought that charge to that point against the force requires work. The electric potential at any point is defined as the energy required to bring a unit test charge from an infinite distance slowly to that point. It is usually measured in volts, and one volt is the potential for which one joule of work must be expended to bring a charge of one coulomb from infinity. This definition of potential, while formal, has little practical application, and a more useful concept is that of electric potential difference, and is the energy required to move a unit charge between two specified points. An electric field has the special property that it is *conservative*, which means that the path taken by the test charge is irrelevant: all paths between two specified points expend the same energy, and thus a unique value for potential difference may be stated. The volt is so strongly identified as the unit of choice for measurement and description of electric potential difference that the term voltage sees greater everyday usage.

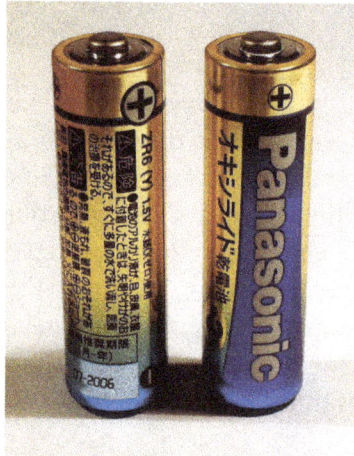

A pair of AA cells. The + sign indicates the polarity of the potential difference between the battery terminals.

For practical purposes, it is useful to define a common reference point to which potentials may be expressed and compared. While this could be at infinity, a much more useful reference is the Earth itself, which is assumed to be at the same potential everywhere. This reference point naturally takes the name earth or ground. Earth is assumed to be an infinite source of equal amounts of positive and negative charge, and is therefore electrically uncharged—and unchargeable.

Electric potential is a scalar quantity, that is, it has only magnitude and not direction. It may be viewed as analogous to height: just as a released object will fall through a difference in heights caused by a gravitational field, so a charge will 'fall' across the voltage caused by an electric field. As relief maps show contour lines marking points of equal height, a set of lines marking points of equal potential (known as equipotentials) may be drawn around an electrostatically charged object. The equipotentials cross all lines of force at right angles. They must also lie parallel to a conductor's surface, otherwise this would produce a force that will move the charge carriers to even the potential of the surface.

The electric field was formally defined as the force exerted per unit charge, but the concept of potential allows for a more useful and equivalent definition: the electric field is the local gradient of the electric potential. Usually expressed in volts per metre, the vector direction of the field is the line of greatest slope of potential, and where the equipotentials lie closest together.

Electromagnets

Ørsted›s discovery in 1821 that a magnetic field existed around all sides of a wire carrying an electric current indicated that there was a direct relationship between electricity and magnetism. Moreover, the interaction seemed different from gravitational and electrostatic forces, the two forces of nature then known. The force on the compass needle did not direct it to or away from the current-carrying wire, but acted at right angles to it. Ørsted's slightly obscure words were that "the electric conflict acts in a revolving manner." The force also depended on the direction of the current, for if the flow was reversed, then the force did too.

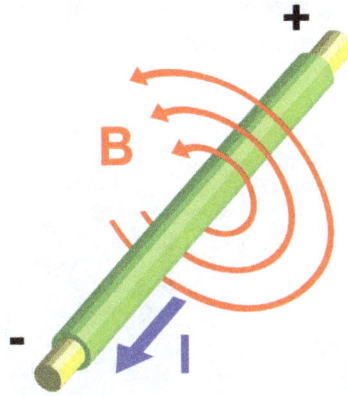

Magnetic field circles around a current

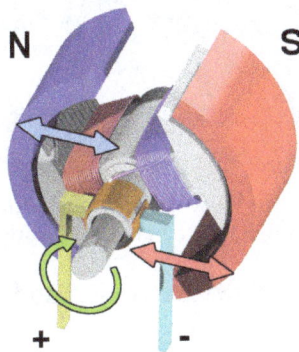

The electric motor exploits an important effect of electromagnetism: a current through a magnetic field experiences a force at right angles to both the field and current.

Ørsted did not fully understand his discovery, but he observed the effect was reciprocal: a current exerts a force on a magnet, and a magnetic field exerts a force on a current. The phenomenon was further investigated by Ampère, who discovered that two parallel current-carrying wires exerted a force upon each other: two wires conducting currents in the same direction are attracted to each other, while wires containing currents in opposite directions are forced apart. The interaction is mediated by the magnetic field each current produces and forms the basis for the international definition of the ampere.

This relationship between magnetic fields and currents is extremely important, for it led to Michael Faraday's invention of the electric motor in 1821. Faraday's homopolar motor consisted of a permanent magnet sitting in a pool of mercury. A current was allowed through a wire suspended from a pivot above the magnet and dipped into the mercury. The magnet exerted a tangential force on the wire, making it circle around the magnet for as long as the current was maintained.

Experimentation by Faraday in 1831 revealed that a wire moving perpendicular to a magnetic field developed a potential difference between its ends. Further analysis of this process, known as electromagnetic induction, enabled him to state the principle, now known as Faraday's law of induction, that the potential difference induced in a closed circuit is proportional to the rate of change of magnetic flux through the loop. Exploitation of this discovery enabled him to invent the first electrical generator in 1831, in which he converted the mechanical energy of a rotating copper disc to electrical energy. Faraday's disc was inefficient and of no use as a practical generator, but it showed the possibility of generating electric power using magnetism, a possibility that would be taken up by those that followed on from his work.

Electrochemistry

Italian physicist Alessandro Volta showing his *"battery"* to French emperor Napoleon Bonaparte in the early 19th century.

The ability of chemical reactions to produce electricity, and conversely the ability of electricity to drive chemical reactions has a wide array of uses.

Electrochemistry has always been an important part of electricity. From the initial invention of the Voltaic pile, electrochemical cells have evolved into the many different types of batteries, electroplating and electrolysis cells. Aluminium is produced in vast quantities this way, and many portable devices are electrically powered using rechargeable cells.

Electric Circuits

An electric circuit is an interconnection of electric components such that electric charge is made to flow along a closed path (a circuit), usually to perform some useful task.

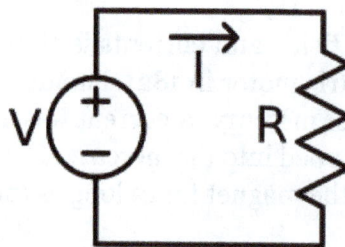

A basic electric circuit. The voltage source V on the left drives a current I around the circuit, delivering electrical energy into the resistor R. From the resistor, the current returns to the source, completing the circuit.

The components in an electric circuit can take many forms, which can include elements such as resistors, capacitors, switches, transformers and electronics. Electronic circuits contain active components, usually semiconductors, and typically exhibit non-linear behaviour, requiring complex analysis. The simplest electric components are those that are termed passive and linear, while they may temporarily store energy, they contain no sources of it, and exhibit linear responses to stimuli.

The resistor is perhaps the simplest of passive circuit elements: as its name suggests, it resists the current through it, dissipating its energy as heat. The resistance is a consequence of the motion of charge through a conductor: in metals, for example, resistance is primarily due to collisions between electrons and ions. Ohm's law is a basic law of circuit theory, stating that the current passing through a resistance is directly proportional to the potential difference across it. The resistance of most materials is relatively constant over a range of temperatures and currents; materials under these conditions are known as 'ohmic'. The ohm, the unit of resistance, was named in honour of Georg Ohm, and is symbolised by the Greek letter Ω. $1\ \Omega$ is the resistance that will produce a potential difference of one volt in response to a current of one amp.

The capacitor is a development of the Leyden jar and is a device that can store charge, and thereby storing electrical energy in the resulting field. It consists of two conducting plates separated by a thin insulating dielectric layer; in practice, thin metal foils are coiled together, increasing the surface area per unit volume and therefore the capacitance. The unit of capacitance is the farad, named after Michael Faraday, and given the symbol F: one farad is the capacitance that develops a potential difference of one volt when it stores a charge of one coulomb. A capacitor connected to a voltage supply initially causes a current as it accumulates charge; this current will however decay in time as the capacitor fills, eventually falling to zero. A capacitor will therefore not permit a steady state current, but instead blocks it.

The inductor is a conductor, usually a coil of wire, that stores energy in a magnetic field in response to the current through it. When the current changes, the magnetic field does too, inducing a voltage between the ends of the conductor. The induced voltage is proportional to the time rate of change of the current. The constant of proportionality is termed the inductance. The unit of inductance is the henry, named after Joseph Henry, a contemporary of Faraday. One henry is the inductance that will induce a potential difference of one volt if the current through it changes at a rate of one ampere per second. The inductor's behaviour is in some regards converse to that of the capacitor: it will freely allow an unchanging current, but opposes a rapidly changing one.

Electric Power

Electric power is the rate at which electric energy is transferred by an electric circuit. The SI unit of power is the watt, one joule per second.

Electric power, like mechanical power, is the rate of doing work, measured in watts, and represented by the letter P. The term *wattage* is used colloquially to mean "electric power in watts." The electric power in watts produced by an electric current I consisting of a charge of Q coulombs every t seconds passing through an electric potential (voltage) difference of V is

$$P = \text{work done per unit time} = \frac{QV}{t} = IV$$

where

Q is electric charge in coulombs

t is time in seconds

I is electric current in amperes

V is electric potential or voltage in volts

Electricity generation is often done with electric generators, but can also be supplied by chemical sources such as electric batteries or by other means from a wide variety of sources of energy. Electric power is generally supplied to businesses and homes by the electric power industry. Electricity is usually sold by the kilowatt hour (3.6 MJ) which is the product of power in kilowatts multiplied by running time in hours. Electric utilities measure power using electricity meters, which keep a running total of the electric energy delivered to a customer. Unlike fossil fuels, electricity is a low entropy form of energy and can be converted into motion or many other forms of energy with high efficiency.

Electronics

Surface mount electronic components

Electronics deals with electrical circuits that involve active electrical components such as vacuum tubes, transistors, diodes, optoelectronics, sensors and integrated circuits, and associated passive interconnection technologies. The nonlinear behaviour of active components and their ability to control electron flows makes amplification of weak signals possible and electronics is widely used in information processing, telecommunications, and signal processing. The ability of electronic devices to act as switches makes digital information processing possible. Interconnection technologies such as circuit boards, electronics packaging technology, and other varied forms of communication infrastructure complete circuit functionality and transform the mixed components into a regular working system.

Today, most electronic devices use semiconductor components to perform electron control. The study of semiconductor devices and related technology is considered a branch of solid state physics, whereas the design and construction of electronic circuits to solve practical problems come under electronics engineering.

Electromagnetic Wave

Faraday's and Ampère's work showed that a time-varying magnetic field acted as a source of an electric field, and a time-varying electric field was a source of a magnetic field. Thus, when either field is changing in time, then a field of the other is necessarily induced. Such a phenomenon has the properties of a wave, and is naturally referred to as an electromagnetic wave. Electromagnetic waves were analysed theoretically by James Clerk Maxwell in 1864. Maxwell developed a set of equations that could unambiguously describe the interrelationship between electric field, magnetic field, electric charge, and electric current. He could moreover prove that such a wave would necessarily travel at the speed of light, and thus light itself was a form of electromagnetic radiation. Maxwell's Laws, which unify light, fields, and charge are one of the great milestones of theoretical physics.

Thus, the work of many researchers enabled the use of electronics to convert signals into high frequency oscillating currents, and via suitably shaped conductors, electricity permits the transmission and reception of these signals via radio waves over very long distances.

Production and uses

Generation and Transmission

In the 6th century BC, the Greek philosopher Thales of Miletus experimented with amber rods and these experiments were the first studies into the production of electrical energy. While this method, now known as the triboelectric effect, can lift light objects and generate sparks, it is extremely inefficient. It was not until the invention of the voltaic pile in the eighteenth century that a viable source of electricity became available. The voltaic pile, and its modern descendant, the electrical battery, store energy chemically and make it available on demand in the form of electrical energy. The battery is a versatile and very common power source which is ideally suited to many applications, but its energy storage is finite, and once discharged it must be disposed of or recharged. For large electrical demands electrical energy must be generated and transmitted continuously over conductive transmission lines.

Early 20th-century alternator made in Budapest, Hungary, in the power generating hall of a hydroelectric station (photograph by Prokudin-Gorsky, 1905–1915).

Electrical power is usually generated by electro-mechanical generators driven by steam produced from fossil fuel combustion, or the heat released from nuclear reactions; or from other sources such as kinetic energy extracted from wind or flowing water. The modern steam turbine invented by Sir Charles Parsons in 1884 today generates about 80 percent of the electric power in the world using a variety of heat sources. Such generators bear no resemblance to Faraday's homopolar disc generator of 1831, but they still rely on his electromagnetic principle that a conductor linking a changing magnetic field induces a potential difference across its ends. The invention in the late nineteenth century of the transformer meant that electrical power could be transmitted more efficiently at a higher voltage but lower current. Efficient electrical transmission meant in turn that electricity could be generated at centralised power stations, where it benefited from economies of scale, and then be despatched relatively long distances to where it was needed.

Wind power is of increasing importance in many countries

Since electrical energy cannot easily be stored in quantities large enough to meet demands on a national scale, at all times exactly as much must be produced as is required. This requires electricity utilities to make careful predictions of their electrical loads, and maintain constant co-ordination with their power stations. A certain amount of generation must always be held in reserve to cushion an electrical grid against inevitable disturbances and losses.

Demand for electricity grows with great rapidity as a nation modernises and its economy develops. The United States showed a 12% increase in demand during each year of the first three decades of the twentieth century, a rate of growth that is now being experienced by emerging economies such as those of India or China. Historically, the growth rate for electricity demand has outstripped that for other forms of energy.

Environmental concerns with electricity generation have led to an increased focus on generation from renewable sources, in particular from wind and hydropower. While debate can be expected to continue over the environmental impact of different means of electricity production, its final form is relatively clean.

Applications

Electricity is a very convenient way to transfer energy, and it has been adapted to a huge, and growing, number of uses. The invention of a practical incandescent light bulb in the 1870s led to lighting becoming one of the first publicly available applications of electrical power. Although electrification brought with it its own dangers, replacing the naked flames of gas lighting greatly reduced fire hazards within homes and factories. Public utilities were set up in many cities targeting the burgeoning market for electrical lighting. In the late 20th century and in modern times, the trend has started to flow in the direction of deregulation in the electrical power sector.

The light bulb, an early application of electricity, operates by Joule heating:
the passage of current through resistance generating heat.

The resistive Joule heating effect employed in filament light bulbs also sees more direct use in electric heating. While this is versatile and controllable, it can be seen as wasteful, since most electrical generation has already required the production of heat at a power station. A number of countries, such as Denmark, have issued legislation restricting or banning the use of resistive electric heating in new buildings. Electricity is however still a highly practical energy source for heating and refrigeration, with air conditioning/heat pumps representing a growing sector for electricity demand for heating and cooling, the effects of which electricity utilities are increasingly obliged to accommodate.

Electricity is used within telecommunications, and indeed the electrical telegraph, demonstrated commercially in 1837 by Cooke and Wheatstone, was one of its earliest applications. With the construction of first intercontinental, and then transatlantic, telegraph systems in the 1860s, electricity had enabled communications in minutes across the globe. Optical fibre and satellite communication have taken a share of the market for communications systems, but electricity can be expected to remain an essential part of the process.

The effects of electromagnetism are most visibly employed in the electric motor, which provides a clean and efficient means of motive power. A stationary motor such as a winch is easily provided with a supply of power, but a motor that moves with its application, such as an electric vehicle, is obliged to either carry along a power source such as a battery, or to collect current from a sliding contact such as a pantograph.

Electronic devices make use of the transistor, perhaps one of the most important inventions of the twentieth century, and a fundamental building block of all modern circuitry. A modern integrated circuit may contain several billion miniaturised transistors in a region only a few centimetres square.

Electricity is also used to fuel public transportation, including electric buses and trains.

Electricity and the Natural World

Physiological Effects

A voltage applied to a human body causes an electric current through the tissues, and although the relationship is non-linear, the greater the voltage, the greater the current. The threshold for perception varies with the supply frequency and with the path of the current, but is about 0.1 mA to 1 mA for mains-frequency electricity, though a current as low as a microamp can be detected as an electrovibration effect under certain conditions. If the current is sufficiently high, it will cause muscle contraction, fibrillation of the heart, and tissue burns. The lack of any visible sign that a conductor is electrified makes electricity a particular hazard. The pain caused by an electric shock can be intense, leading electricity at times to be employed as a method of torture. Death caused by an electric shock is referred to as electrocution. Electrocution is still the means of judicial execution in some jurisdictions, though its use has become rarer in recent times.

Electrical Phenomena in Nature

The electric eel, *Electrophorus electricus*

Electricity is not a human invention, and may be observed in several forms in nature, a prominent manifestation of which is lightning. Many interactions familiar at the macroscopic level, such as touch, friction or chemical bonding, are due to interactions between electric fields on the atomic scale. The Earth's magnetic field is thought to arise from a natural dynamo of circulating currents in the planet's core. Certain crystals, such as quartz, or even sugar, generate a potential difference across their faces when subjected to external pressure. This phenomenon is known as piezoelectricity, from the Greek *piezein*, meaning to press, and was discovered in 1880 by Pierre and Jacques Curie. The effect is reciprocal, and when a piezoelectric material is subjected to an electric field, a small change in physical dimensions takes place.

Some organisms, such as sharks, are able to detect and respond to changes in electric fields, an ability known as electroreception, while others, termed electrogenic, are able to generate voltages themselves to serve as a predatory or defensive weapon. The order Gymnotiformes, of which the best known example is the electric eel, detect or stun their prey via high voltages generated from modified muscle cells called electrocytes. All animals transmit information along their cell membranes with voltage pulses called action potentials, whose functions include communication by the nervous system between neurons and muscles. An electric shock stimulates this system, and causes muscles to contract. Action potentials are also responsible for coordinating activities in certain plants.

Cultural Perception

In 1850, William Gladstone asked the scientist Michael Faraday why electricity was valuable. Faraday answered, "One day sir, you may tax it."

In the 19th and early 20th century, electricity was not part of the everyday life of many people, even in the industrialised Western world. The popular culture of the time accordingly often depicts it as a mysterious, quasi-magical force that can slay the living, revive the dead or otherwise bend the laws of nature. This attitude began with the 1771 experiments of Luigi Galvani in which the legs of dead frogs were shown to twitch on application of animal electricity. "Revitalization" or resuscitation of apparently dead or drowned persons was reported in the medical literature shortly after Galvani's work. These results were known to Mary Shelley when she authored *Frankenstein* (1819), although she does not name the method of revitalization of the monster. The revitalization of monsters with electricity later became a stock theme in horror films.

As the public familiarity with electricity as the lifeblood of the Second Industrial Revolution grew, its wielders were more often cast in a positive light, such as the workers who "finger death at their gloves' end as they piece and repiece the living wires" in Rudyard Kipling's 1907 poem *Sons of Martha*. Electrically powered vehicles of every sort featured large in adventure stories such as those of Jules Verne and the *Tom Swift* books. The masters of electricity, whether fictional or real—including scientists such as Thomas Edison, Charles Steinmetz or Nikola Tesla—were popularly conceived of as having wizard-like powers.

With electricity ceasing to be a novelty and becoming a necessity of everyday life in the later half of the 20th century, it required particular attention by popular culture only when it *stops* flowing, an event that usually signals disaster. The people who *keep* it flowing, such as the nameless hero of Jimmy Webb's song "Wichita Lineman" (1968), are still often cast as heroic, wizard-like figures.

Electric Charge

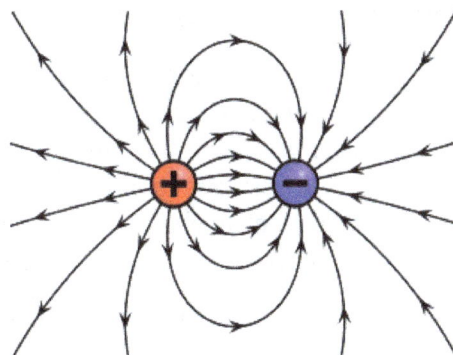

Electric field of a positive and a negative point charge

Electric charge is the physical property of matter that causes it to experience a force when placed in an electromagnetic field. There are two types of electric charges: *positive* and *negative* (commonly carried by protons and electrons respectively). Like charges repel and unlike attract. An absence of net charge is referred to as *neutral*. An object is negatively charged if it has an excess of electrons, and is otherwise positively charged or uncharged. The SI derived unit of electric charge is the coulomb (C). In electrical engineering, it is also common to use the ampere-hour (Ah), and, in chemistry, it is common to use the elementary charge (*e*) as a unit. The symbol *Q* often denotes charge. Early knowledge of how charged substances interact is now called classical electrodynamics, and is still accurate for problems that don't require consideration of quantum effects.

The *electric charge* is a fundamental conserved property of some subatomic particles, which determines their electromagnetic interaction. Electrically charged matter is influenced by, and produces, electromagnetic fields. The interaction between a moving charge and an electromagnetic field is the source of the electromagnetic force, which is one of the four fundamental forces.

Twentieth-century experiments demonstrated that electric charge is *quantized*; that is, it comes in integer multiples of individual small units called the elementary charge, *e*, approximately equal to 1.602×10^{-19} coulombs (except for particles called quarks, which have charges that are integer multiples of $\frac{1}{3}$ e). The proton has a charge of +*e*, and the electron has a charge of −*e*. The study of charged particles, and how their interactions are mediated by photons, is called quantum electrodynamics.

Overview

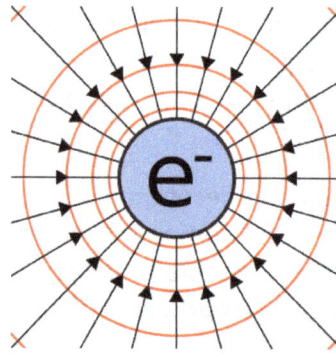

Diagram showing field lines and equipotentials around an electron, a negatively charged particle.
In an electrically neutral atom, the number of electrons is equal to the number of protons
(which are positively charged), resulting in a net zero overall charge.

Charge is the fundamental property of forms of matter that exhibit electrostatic attraction or repulsion in the presence of other matter. Electric charge is a characteristic property of many subatomic particles. The charges of free-standing particles are integer multiples of the elementary charge *e*; we say that electric charge is *quantized*. Michael Faraday, in his electrolysis experiments, was the first to note the discrete nature of electric charge. Robert Millikan's oil drop experiment demonstrated this fact directly, and measured the elementary charge.

By convention, the charge of an electron is −1, while that of a proton is +1. Charged particles whose charges have the same sign repel one another, and particles whose charges have different signs attract. Coulomb's law quantifies the electrostatic force between two particles by asserting that the force is proportional to the product of their charges, and inversely proportional to the square of the distance between them.

The charge of an antiparticle equals that of the corresponding particle, but with opposite sign. Quarks have fractional charges of either $-\frac{1}{3}$ or $+\frac{2}{3}$, but free-standing quarks have never been observed (the theoretical reason for this fact is asymptotic freedom).

The electric charge of a macroscopic object is the sum of the electric charges of the particles that make it up. This charge is often small, because matter is made of atoms, and atoms typically have equal numbers of protons and electrons, in which case their charges cancel out, yielding a net charge of zero, thus making the atom neutral.

An *ion* is an atom (or group of atoms) that has lost one or more electrons, giving it a net positive charge (cation), or that has gained one or more electrons, giving it a net negative charge (anion). *Monatomic ions* are formed from single atoms, while *polyatomic ions* are formed from two or more atoms that have been bonded together, in each case yielding an ion with a positive or negative net charge.

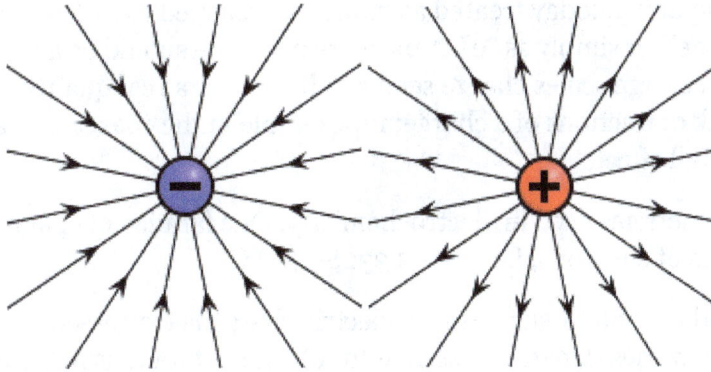

Electric field induced by a positive electric charge (left) and a field induced by a negative electric charge (right).

During formation of macroscopic objects, constituent atoms and ions usually combine to form structures composed of neutral *ionic compounds* electrically bound to neutral atoms. Thus macroscopic objects tend toward being neutral overall, but macroscopic objects are rarely perfectly net neutral.

Sometimes macroscopic objects contain ions distributed throughout the material, rigidly bound in place, giving an overall net positive or negative charge to the object. Also, macroscopic objects made of conductive elements, can more or less easily (depending on the element) take on or give off electrons, and then maintain a net negative or positive charge indefinitely. When the net electric charge of an object is non-zero and motionless, the phenomenon is known as static electricity. This can easily be produced by rubbing two dissimilar materials together, such as rubbing amber with fur or glass with silk. In this way non-conductive materials can be charged to a significant degree, either positively or negatively. Charge taken from one material is moved to the other material, leaving an opposite charge of the same magnitude behind. The law of *conservation of charge* always applies, giving the object from which a negative charge is taken a positive charge of the same magnitude, and vice versa.

Even when an object's net charge is zero, charge can be distributed non-uniformly in the object (e.g., due to an external electromagnetic field, or bound polar molecules). In such cases the object is said to be polarized. The charge due to polarization is known as bound charge, while charge on an object produced by electrons gained or lost from outside the object is called *free charge*. The motion of electrons in conductive metals in a specific direction is known as electric current.

Units

The SI unit of quantity of electric charge is the coulomb, which is equivalent to about $6.242×10^{18}$ e (e is the charge of a proton). Hence, the charge of an electron is approximately $-1.602×10^{-19}$ C. The coulomb is defined as the quantity of charge that has passed through the cross section of an electrical conductor carrying one ampere within one second. The symbol Q is often used to denote a quantity of electricity or charge. The quantity of electric charge can be directly measured with an electrometer, or indirectly measured with a ballistic galvanometer.

After finding the quantized character of charge, in 1891 George Stoney proposed the unit 'electron' for this fundamental unit of electrical charge. This was before the discovery of the particle by J.J. Thomson in 1897. The unit is today treated as nameless, referred to as "elementary charge", "fundamental unit of charge", or simply as "e". A measure of charge should be a multiple of the elementary charge e, even if at large scales charge seems to behave as a real quantity. In some contexts it is meaningful to speak of fractions of a charge; for example in the charging of a capacitor, or in the fractional quantum Hall effect.

The unit faraday is sometimes used in electrochemistry. One faraday of charge is the magnitude of the charge of one mole of electrons, i.e. 96485.33289(59) C.

In systems of units other than SI such as cgs, electric charge is expressed as combination of only three fundamental quantities (length, mass, and time), and not four, as in SI, where electric charge is a combination of length, mass, time, and electric current.

History

As reported by the ancient Greek mathematician Thales of Miletus around 600 BC, charge (or *electricity*) could be accumulated by rubbing fur on various substances, such as amber. The Greeks noted that the charged amber buttons could attract light objects such as hair. They also noted that if they rubbed the amber for long enough, they could even get an electric spark to jump. This property derives from the triboelectric effect.

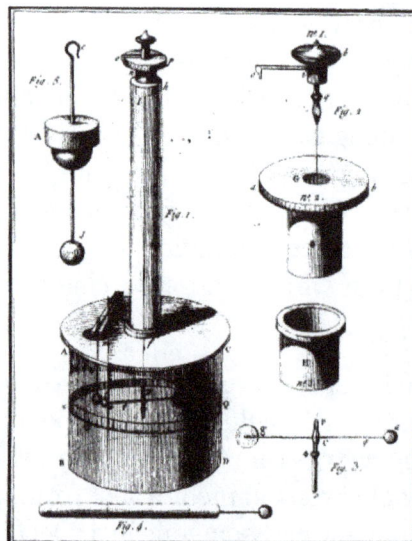

Coulomb's torsion balance

In 1600, the English scientist William Gilbert returned to the subject in *De Magnete*, and coined the New Latin word *electricus* from (ēlektron), the Greek word for *amber*, which soon gave rise to the English words "electric" and "electricity". He was followed in 1660 by Otto von Guericke, who invented what was probably the first electrostatic generator. Other European pioneers were Robert Boyle, who in 1675 stated that electric attraction and repulsion can act across a vacuum; Stephen Gray, who in 1729 classified materials as conductors and insulators; and C. F. du Fay, who proposed in 1733 that electricity comes in two varieties that cancel each other, and expressed this in terms of a two-fluid theory. When glass was rubbed with silk, du Fay said that the glass was charged with *vitreous electricity*, and, when amber was rubbed with fur, the amber was charged with *resinous electricity*. In 1839, Michael Faraday showed that the apparent division between static electricity, current electricity, and bioelectricity was incorrect, and all were a consequence of the behavior of a single kind of electricity appearing in opposite polarities. It is arbitrary which polarity is called positive and which is called negative. Positive charge can be defined as the charge left on a glass rod after being rubbed with silk.

One of the foremost experts on electricity in the 18th century was Benjamin Franklin, who argued in favour of a one-fluid theory of electricity. Franklin imagined electricity as being a type of invisible fluid present in all matter; for example, he believed that it was the glass in a Leyden jar that held the accumulated charge. He posited that rubbing insulating surfaces together caused this fluid to change location, and that a flow of this fluid constitutes an electric current. He also posited that when matter contained too little of the fluid it was "negatively" charged, and when it had an excess it was "positively" charged. For a reason that was not recorded, he identified the term "positive" with vitreous electricity and "negative" with resinous electricity. William Watson arrived at the same explanation at about the same time.

Static Electricity and Electric Current

Static electricity and electric current are two separate phenomena. They both involve electric charge, and may occur simultaneously in the same object. Static electricity refers to the electric charge of an object and the related electrostatic discharge when two objects are brought together that are not at equilibrium. An electrostatic discharge creates a change in the charge of each of the two objects. In contrast, electric current is the flow of electric charge through an object, which produces no net loss or gain of electric charge.

Electrification by Friction

When a piece of glass and a piece of resin—neither of which exhibit any electrical properties—are rubbed together and left with the rubbed surfaces in contact, they still exhibit no electrical properties. When separated, they attract each other.

A second piece of glass rubbed with a second piece of resin, then separated and suspended near the former pieces of glass and resin causes these phenomena:

- The two pieces of glass repel each other.

- Each piece of glass attracts each piece of resin.

- The two pieces of resin repel each other.

This attraction and repulsion is an *electrical phenomena*, and the bodies that exhibit them are said to be *electrified*, or *electrically charged*. Bodies may be electrified in many other ways, as well as by friction. The electrical properties of the two pieces of glass are similar to each other but opposite to those of the two pieces of resin: The glass attracts what the resin repels and repels what the resin attracts.

If a body electrified in any manner whatsoever behaves as the glass does, that is, if it repels the glass and attracts the resin, the body is said to be *vitreously* electrified, and if it attracts the glass and repels the resin it is said to be *resinously* electrified. All electrified bodies are either vitreously or resinously electrified.

An established convention in the scientific community defines vitreous electrification as positive, and resinous electrification as negative. The exactly opposite properties of the two kinds of electrification justify our indicating them by opposite signs, but the application of the positive sign to one rather than to the other kind must be considered as a matter of arbitrary convention—just as it is a matter of convention in mathematical diagram to reckon positive distances towards the right hand.

No force, either of attraction or of repulsion, can be observed between an electrified body and a body not electrified.

Actually, all bodies are electrified, but may appear not electrified because of the relatively similar charge of neighboring objects in the environment. An object further electrified + or − creates an equivalent or opposite charge by default in neighboring objects, until those charges can equalize. The effects of attraction can be observed in high-voltage experiments, while lower voltage effects are merely weaker and therefore less obvious. The attraction and repulsion forces are codified by Coulomb's law (attraction falls off at the square of the distance, which has a corollary for acceleration in a gravitational field, suggesting that gravitation may be merely electrostatic phenomenon between relatively weak charges in terms of scale).

It is now known that the Franklin-Watson model was fundamentally correct. There is only one kind of electrical charge, and only one variable is required to keep track of the amount of charge. On the other hand, just knowing the charge is not a complete description of the situation. Matter is composed of several kinds of electrically charged particles, and these particles have many properties, not just charge.

The most common charge carriers are the positively charged proton and the negatively charged electron. The movement of any of these charged particles constitutes an electric current. In many situations, it suffices to speak of the *conventional current* without regard to whether it is carried by positive charges moving in the direction of the conventional current or by negative charges moving in the opposite direction. This macroscopic viewpoint is an approximation that simplifies electromagnetic concepts and calculations.

At the opposite extreme, if one looks at the microscopic situation, one sees there are many ways of carrying an electric current, including: a flow of electrons; a flow of electron "holes" that act like positive particles; and both negative and positive particles (ions or other charged particles) flowing in opposite directions in an electrolytic solution or a plasma.

Beware that, in the common and important case of metallic wires, the direction of the conventional current is opposite to the drift velocity of the actual charge carriers; i.e., the electrons. This is a source of confusion for beginners.

Conservation of Electric Charge

The total electric charge of an isolated system remains constant regardless of changes within the system itself. This law is inherent to all processes known to physics and can be derived in a local form from gauge invariance of the wave function. The conservation of charge results in the charge-current continuity equation. More generally, the net change in charge density ρ within a volume of integration V is equal to the area integral over the current density J through the closed surface $S = \partial V$, which is in turn equal to the net current I:

$$-\frac{d}{dt}\int_V \rho dV = \oiint_{av} J.dS = \int JdS \cos\theta = I.$$

Thus, the conservation of electric charge, as expressed by the continuity equation, gives the result:

$$I = -\frac{dQ}{dt}.$$

The charge transferred between times t_i and t_f is obtained by integrating both sides:

$$Q = \int_{t_i}^{t_f} Idt$$

where I is the net outward current through a closed surface and Q is the electric charge contained within the volume defined by the surface.

Relativistic Invariance

Charge is a relativistic invariant. This means that any particle that has charge Q, no matter how fast it goes, always has charge Q. This property has been experimentally verified by showing that the charge of *one* helium nucleus (two protons and two neutrons bound together in a nucleus and moving around at high speeds) is the same as *two* deuterium nuclei (one proton and one neutron bound together, but moving much more slowly than they would if they were in a helium nucleus).

Electric Field and Dielectrics

Dielectrics are material that can store electrostatic charge. The principles of dielectrics are fundamental in the development of the capacitor. Dielectrics are an emerging field of study; the following section will not only provide an overview, it will also delve deep into the variegated topics related to it.

Electric Field

An electric field is a vector field that associates to each point in space the Coulomb force that would be experienced per unit of electric charge, by an infinitesimal test charge at that point. Electric fields are created by electric charges and can be induced by time-varying magnetic fields. The electric field combines with the magnetic field to form the electromagnetic field.

Definition

The electric field, E, at a given point is defined as the (vector) force, F, that would be exerted on a stationary test particle of unit charge by electromagnetic forces (i.e. the Lorentz force). A particle of charge q would be subject to a force F=qE.

Its SI units are newtons per coulomb (N·C⁻¹) or, equivalently, volts per metre (V·m⁻¹), which in terms of SI base units are kg·m·s⁻³·A⁻¹.

Sources of Electric Field

Causes and Description

Electric fields are caused by electric charges or varying magnetic fields. The former effect is described by Gauss's law, the latter by Faraday's law of induction, which together are enough to define the behavior of the electric field as a function of charge repartition and magnetic field. However, since the magnetic field is described as a function of electric field, the equations of both fields are coupled and together form Maxwell's equations that describe both fields as a function of charges and currents.

In the special case of a steady state (stationary charges and currents), the Maxwell-Faraday inductive effect disappears. The resulting two equations (Gauss's law $\nabla \cdot \mathbf{E} = \dfrac{\rho}{\varepsilon_0}$ and Faraday's law with no induction term $\nabla \times \mathbf{E} = 0$), taken together, are equivalent to Coulomb's law, written as

$$\mathbf{E}(\mathbf{r}) = \frac{1}{4\pi\varepsilon_0} \int \rho(\mathbf{r}') \frac{\mathbf{r} - \mathbf{r}'}{|\mathbf{r} - \mathbf{r}'|^3} d^3 r'$$ for a charge density $\rho(\mathbf{r})$ (r denotes the position in space). Notice

that ε_0, the permittivity of vacuum, must be substituted if charges are considered in non-empty media.

Continuous Vs. Discrete Charge Representation

The equations of electromagnetism are best described in a continuous description. However, charges are sometimes best described as discrete points; for example, some models may describe electrons as point sources where charge density is infinite on an infinitesimal section of space.

A charge q located at r_0 can be described mathematically as a charge density $\rho(r) = q\delta(r - r_0)$, where the Dirac delta function (in three dimensions) is used. Conversely, a charge distribution can be approximated by many small point charges.

Superposition Principle

Electric fields satisfy the superposition principle, because Maxwell's equations are linear. As a result, if E_1 and E_2 are the electric fields resulting from distribution of charges ρ_1 and ρ_2, a distribution of charges $\rho_1 + \rho_2$ will create an electric field $E_1 + E_2$; for instance, Coulomb's law is linear in charge density as well.

This principle is useful to calculate the field created by multiple point charges. If charges $q_1, q_2, ..., q_n$ are stationary in space at $r_1, r_2, ...r_n$, in the absence of currents, the superposition principle proves that the resulting field is the sum of fields generated by each particle as described by Coulomb's law:

$$\mathbf{E}(\mathbf{r}) = \sum_{i=1}^{N} \mathbf{E}_i(\mathbf{r}) = \frac{1}{4\pi\varepsilon_0} \sum_{i=1}^{N} q_i \frac{\mathbf{r} - \mathbf{r}_i}{|\mathbf{r} - \mathbf{r}_i|^3}$$

Electrostatic Fields

Electrostatic fields are E-fields which do not change with time, which happens when charges and currents are stationary. In that case, Coulomb's law fully describes the field.

Electric Potential

If a system is static, such that magnetic fields are not time-varying, then by Faraday's law, the electric field is curl-free. In this case, one can define an electric potential, that is, a function Φ such that $\mathbf{E} = -\nabla\Phi$. This is analogous to the gravitational potential.

Parallels between Electrostatic and Gravitational Fields

Coulomb's law, which describes the interaction of electric charges:

$$\mathbf{F} = q\left(\frac{Q}{4\pi\varepsilon_0} \frac{\hat{\mathbf{r}}}{|\mathbf{r}|^2}\right) = q\mathbf{E}$$

is similar to Newton's law of universal gravitation:

$$\mathbf{F} = m\left(-GM\frac{\hat{\mathbf{r}}}{|\mathbf{r}|^2}\right) = m\mathbf{g}$$

(where $\hat{\mathbf{r}} = \dfrac{\mathbf{r}}{|\mathbf{r}|}$).

This suggests similarities between the electric field E and the gravitational field g, or their associated potentials. Mass is sometimes called "gravitational charge" because of that similarity.

Electrostatic and gravitational forces both are central, conservative and obey an inverse-square law.

Uniform Fields

A uniform field is one in which the electric field is constant at every point. It can be approximated by placing two conducting plates parallel to each other and maintaining a voltage (potential difference) between them; it is only an approximation because of boundary effects (near the edge of the planes, electric field is distorted because the plane does not continue). Assuming infinite planes, the magnitude of the electric field E is:

$$E = -\frac{\Delta\phi}{d}$$

where $\Delta\phi$ is the potential difference between the plates and d is the distance separating the plates. The negative sign arises as positive charges repel, so a positive charge will experience a force away from the positively charged plate, in the opposite direction to that in which the voltage increases. In micro- and nano-applications, for instance in relation to semiconductors, a typical magnitude of an electric field is in the order of 10^6 V·m^{-1}, achieved by applying a voltage of the order of 1 volt between conductors spaced 1 μm apart.

Electrodynamic Fields

Electrodynamic fields are E-fields which do change with time, for instance when charges are in motion.

The electric field cannot be described independently of the magnetic field in that case. If A is the magnetic vector potential, defined so that $\mathbf{B} = \nabla \times \mathbf{A}$, one can still define an electric potential Φ such that:

$$\mathbf{E} = -\nabla\Phi - \frac{\partial\mathbf{A}}{\partial t}$$

One can recover Faraday's law of induction by taking the curl of that equation

$$\nabla \times \mathbf{E} = -\frac{\partial(\nabla \times \mathbf{A})}{\partial t} = -\frac{\partial\mathbf{B}}{\partial t}$$

which justifies, a posteriori, the previous form for E.

Energy in the Electric Field

The total energy per unit volume stored by the electromagnetic field is

$$u_{EM} = \frac{\varepsilon}{2}|\mathbf{E}|^2 + \frac{1}{2\mu}|\mathbf{B}|^2$$

where ε is the permittivity of the medium in which the field exists, μ its magnetic permeability, and E and B are the electric and magnetic field vectors.

As E and B fields are coupled, it would be misleading to split this expression into "electric" and "magnetic" contributions. However, in the steady-state case, the fields are no longer coupled. It makes sense in that case to compute the electrostatic energy per unit volume:

$$u_{ES} = \frac{1}{2}\varepsilon|\mathbf{E}|^2,$$

The total energy U stored in the electric field in a given volume V is therefore

$$U_{ES} = \frac{1}{2}\varepsilon\int_V|\mathbf{E}|^2\,dV,$$

Further Extensions

Definitive Equation of Vector Fields

In the presence of matter, it is helpful to extend the notion of the electric field into three vector fields:

$$\mathbf{D} = \varepsilon_0\mathbf{E} + \mathbf{P}$$

where P is the electric polarization – the volume density of electric dipole moments, and D is the electric displacement field. Since E and P are defined separately, this equation can be used to define D. The physical interpretation of D is not as clear as E (effectively the field applied to the material) or P (induced field due to the dipoles in the material), but still serves as a convenient mathematical simplification, since Maxwell's equations can be simplified in terms of free charges and currents.

Constitutive Relation

The E and D fields are related by the permittivity of the material, ε.

For linear, homogeneous, isotropic materials E and D are proportional and constant throughout the region, there is no position dependence: For inhomogeneous materials, there is a position dependence throughout the material:

$$\mathbf{D}(\mathbf{r}) = \varepsilon\mathbf{E}(\mathbf{r})$$

For anisotropic materials the E and D fields are not parallel, and so E and D are related by the permittivity tensor (a 2nd order tensor field), in component form:

$$D_i = \varepsilon_{ij}E_j$$

For non-linear media, E and D are not proportional. Materials can have varying extents of linearity, homogeneity and isotropy.

Classification of Electric Fields

Electric field configurations can be classified into two forms of fields;

1. Uniform field.

2. Non-Uniform field.

Non-uniform fields can be further distinguished between two types;

1. Weakly nonuniform fields.

2. Extremely nonuniform fields

Electric Fields

Uniform Field
$(U_i = U_b \quad \eta = 1)$
No PB before breakdown

Nonuniform Fields
$(\eta < 1)$

Weakly nonuniform
$(U_i = U_b)(\eta \geq 0.25 \quad$ for atmospheric air)
No PB before breakdown

Extremely nonuniform
$(U_i << U_b)(\eta < 0.25)$
Stable PB before breakdown

PB – Partial Breakdown
U_i – PB inception voltage
U_b – complete breakdown voltage
η – Schwaiger Factor

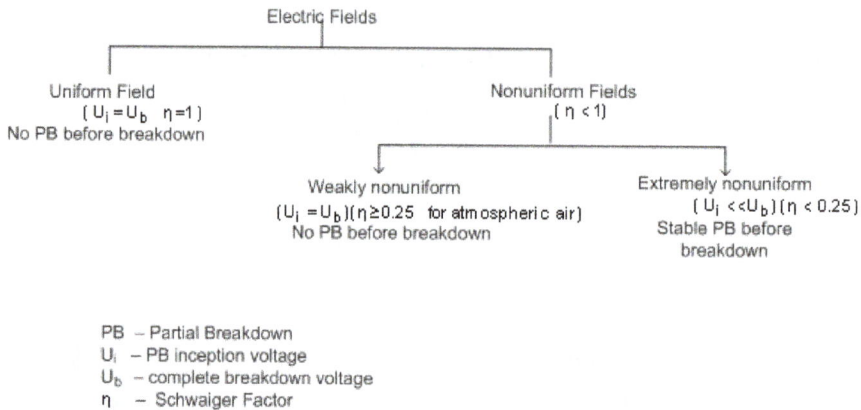

- The two extreme forms of fields, the 'uniform' and the 'extremely nonuniform' fields, are shown in the figure a.

  The electric field between two electrodes is plotted beginning with the equipotential lines. An electrode is always at equi-potential. Hence, the electrode surface is the first equipotential line to be plotted. The field lines intersect the equipotential lines at right angle. Smaller are the squares formed, higher the field intensity.

- In a 'uniform field', the potential is linearly distributed and the electric field intensity is constant throughout the space in the main field region between the two electrodes, as shown in the figure below a(i). An important characteristic of this type of field is that the insulation breakdown takes place without any partial discharge preceding the breakdown ($U_i=U_b$).

- There is an extreme nonlinear distribution of potential in the space between two needle electrodes, leading to strong nonuniformity in electric field intensity as in the figure below a(ii). This is a typical case of an 'extremely nonuniform field'. Unlike in uniform fields, the insulation breakdown in extremely nonuniform fields takes place after stable partial breakdown are set ($U_i<<U_b$). The partial breakdown are rendered to be unstable only just before the complete breakdown. This field configuration is of great technical importance, as it is the most unfavourable condition of electric field faced by a dielectric. At the tip of the elec-

trodes the dielectric is subjected to a very high electric stress, but elsewhere between the electrodes it is stressed moderately.

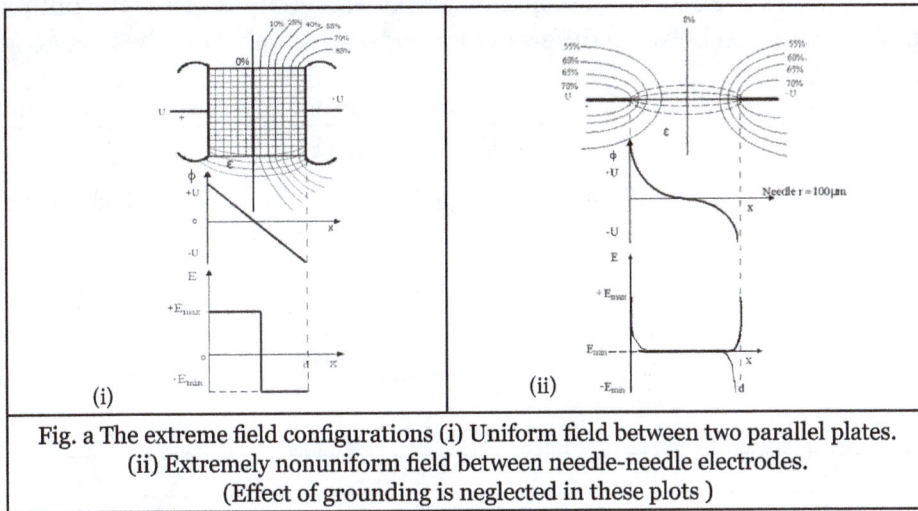

Fig. a The extreme field configurations (i) Uniform field between two parallel plates.
(ii) Extremely nonuniform field between needle-needle electrodes.
(Effect of grounding is neglected in these plots)

- For a given electrode gap distance in uniform field a dielectric has the highest breakdown strength. However, it is very difficult to realize a uniform field in practice. It is accomplished only for experimental purposes in research laboratories with tremendous effort and utmost care. The size of the electrodes may have to be increased extraordinarily large, whereas the slightest irregularity on the electrode surface may change the field characteristics in case of small gap distance.

- Between the two extreme field configurations explained above, another important type of field is classified as 'weakly nonuniform field'. Like in uniform field, in weakly nonuniform fields also no stable partial breakdown occur before the breakdown ($U_i = U_b$). Electrodes like concentric spheres and coaxial cylinders having a 'radial field' are typical examples of weakly nonuniform fields, if the concentric electrode dimensions are suitably designed. The exact value of η, defining weakly nonuniform field, depends upon the particular dielectric and its physical conditions. Nevertheless, the main criterion must be fulfilled that no stable partial breakdown occur before the breakdown.

Degree of Uniformity of Electric Fields

- The degree of uniformity η introduced by Schwaiger in 1922 as a measure of the uniformity of a field, is defined as following

$$\eta = \frac{\hat{E}_{mean}}{\hat{E}_{max}} = \frac{\hat{U}}{d} \cdot \frac{1}{\hat{E}_{max}}$$

or

$$\hat{U} = \hat{E}_{max} \cdot \eta \cdot d$$

$\hat{E}_{mean}$ and $\hat{E}_{max}$ are the peak values of the Mean and the Maximum field Intensities in a dielectric respectively. $\hat{U}$ is the peak value of potential difference applied between the two electrodes at a distance 'd' apart.

- The value of η also represents the degree of utilization of the dielectric in between two electrodes. A higher value of η represents better utilization of the insulating properties of a dielectric. Thus η, a dimensionless quantity enables a comparison of the uniformity of field configuration formed between different electrodes. The value of η lies between, $0 \leq \eta \leq 1$.

Field Classification	Uniform	Weakly non-uniform	Extremely non-uniform
Electrode Confi uration	Parallel plates	Concentric cylinders $r_i = 0.25\, r_o$	Needle-plane
η	1	≤ 0.25	<< 0.01

- Knowledge of the value for η serves as a ready reference, an important information for insulation design in equipment. For determining the exact magnitude of maximum electric stress, numerical estimation techniques have to be applied for the shapes of electrodes used in the equipment.

- Schwaiger also introduced 'p', a geometrical characteristics for an electrode configuration and established that it is possible to represent η as a function of 'p',

$$p = \frac{r+d}{r}$$

and $\quad (1 \leq p < \infty)$

$$\eta = f(p)$$

where r is the radius of curvature of the sharpest electrode and d the shortest gap distance between the two electrodes under consideration.

For some common and practical electrode configurations, the equation is represented graphically in Figure b in double logarithmic scale. These are known as 'Schwaiger curves'.

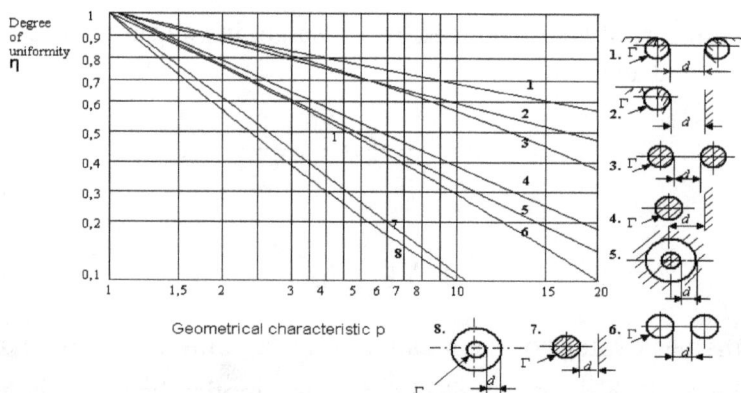

Fig. b Schwaiger curves for spherical, cylindrical and curved electrode field configurations

For a fixed value of 'p', the following important basic relations of dependency of η are observed from these curves,

- fields between cylindrical electrode systems; cylinder-cylinder (3), cylinder - plane (4), concentric cylinders (5) etc., have a higher value of η, that is, they are more uniform than the fields in spherical electrode systems; sphere - sphere (6), sphere - plane (7), concentric spheres (8), etc.

- a symmetrical electrode system, for example, sphere - sphere or cylinder - cylinder, has a higher value of η than the corresponding unsymmetrical system, that is sphere - plane or cylinder - plane systems.

- the field between two similar electrodes, cylinders or spheres, placed adjacent to each other is more uniform or has a higher value of η than when the electrodes are placed coaxial or concentric.

Dielectric

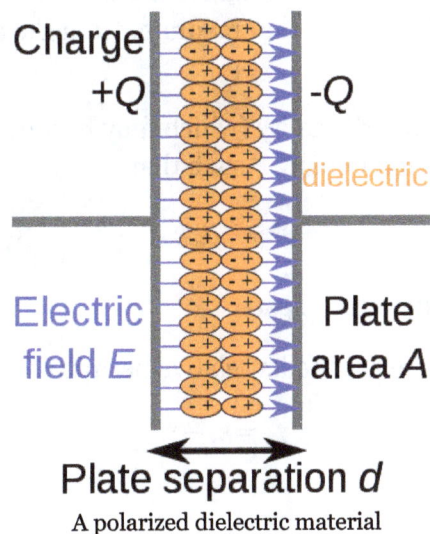

A polarized dielectric material

A dielectric (or dielectric material) is an electrical insulator that can be polarized by an applied electric field. When a dielectric is placed in an electric field, electric charges do not flow through the material as they do in a electrical conductor, but only slightly shift from their average equilibrium positions causing dielectric polarization. Because of dielectric polarization, positive charges are displaced toward the field and negative charges shift in the opposite direction. This creates an internal electric field that reduces the overall field within the dielectric itself. If a dielectric is composed of weakly bonded molecules, those molecules not only become polarized, but also reorient so that their symmetry axes align to the field.

The study of dielectric properties concerns storage and dissipation of electric and magnetic energy in materials. Dielectrics are important for explaining various phenomena in electronics, optics, solid-state physics, and cell biophysics.

Terminology

Although the term *insulator* implies low electrical conduction, *dielectric* typically means materials with a high polarizability. The latter is expressed by a number called the relative permittivity (also known in older texts as dielectric constant). The term insulator is generally used to indicate electrical obstruction while the term dielectric is used to indicate the energy storing capacity of the material (by means of polarization). A common example of a dielectric is the electrically insulating material between the metallic plates of a capacitor. The polarization of the dielectric by the applied electric field increases the capacitor's surface charge for the given electric field strength.

The term "dielectric" was coined by William Whewell (from "dia-electric") in response to a request from Michael Faraday. A *perfect dielectric* is a material with zero electrical conductivity (cf. perfect conductor), thus exhibiting only a displacement current; therefore it stores and returns electrical energy as if it were an ideal capacitor.

Electric Susceptibility

The electric susceptibility χ_e of a dielectric material is a measure of how easily it polarizes in response to an electric field. This, in turn, determines the electric permittivity of the material and thus influences many other phenomena in that medium, from the capacitance of capacitors to the speed of light.

It is defined as the constant of proportionality (which may be a tensor) relating an electric field E to the induced dielectric polarization density P such that

$$\mathbf{P} = \varepsilon_0 \chi_e \mathbf{E},$$

where ε_0 is the electric permittivity of free space.

The susceptibility of a medium is related to its relative permittivity ε_r by

$$\chi_e = \varepsilon_r - 1.$$

So in the case of a vacuum,

$$\chi_e = 0.$$

The electric displacement D is related to the polarization density P by

$$\mathbf{D} = \varepsilon_0 \mathbf{E} + \mathbf{P} = \varepsilon_0 (1 + \chi_e)\mathbf{E} = \varepsilon_r \varepsilon_0 \mathbf{E}.$$

Dispersion and Causality

In general, a material cannot polarize instantaneously in response to an applied field. The more general formulation as a function of time is

$$\mathbf{P}(t) = \varepsilon_0 \int_{-\infty}^{t} \chi_e (t - t')\mathbf{E}(t')dt'.$$

That is, the polarization is a convolution of the electric field at previous times with time-dependent susceptibility given by $\chi_e(\Delta t)$. The upper limit of this integral can be extended to infinity as well if one defines $\chi_e(\Delta t) = 0$ for $\Delta t < 0$. An instantaneous response corresponds to Dirac delta function susceptibility $\chi_e(\Delta t) = \chi_e \delta(\Delta t)$.

It is more convenient in a linear system to take the Fourier transform and write this relationship as a function of frequency. Due to the convolution theorem, the integral becomes a simple product,

$$\mathbf{P}(\omega) = \varepsilon_0 \chi_e(\omega) \mathbf{E}(\omega).$$

Note the simple frequency dependence of the susceptibility, or equivalently the permittivity. The shape of the susceptibility with respect to frequency characterizes the dispersion properties of the material.

Moreover, the fact that the polarization can only depend on the electric field at previous times (i.e., $\chi_e(\Delta t) = 0$ for $\Delta t < 0$, a consequence of causality, imposes Kramers–Kronig constraints on the real and imaginary parts of the susceptibility $\chi_e(\omega)$.

Dielectric Polarization

Basic Atomic Model

In the classical approach to the dielectric model, a material is made up of atoms. Each atom consists of a cloud of negative charge (electrons) bound to and surrounding a positive point charge at its center. In the presence of an electric field the charge cloud is distorted, as shown in the top right of the figure.

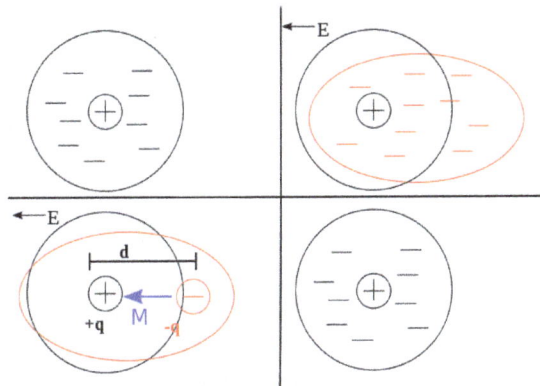

Electric field interaction with an atom under the classical dielectric model.

This can be reduced to a simple dipole using the superposition principle. A dipole is characterized by its dipole moment, a vector quantity shown in the figure as the blue arrow labeled M. It is the relationship between the electric field and the dipole moment that gives rise to the behavior of the dielectric. (Note that the dipole moment points in the same direction as the electric field in the figure. This isn't always the case, and is a major simplification, but is true for many materials.)

When the electric field is removed the atom returns to its original state. The time required to do so is the so-called relaxation time; an exponential decay.

This is the essence of the model in physics. The behavior of the dielectric now depends on the situation. The more complicated the situation, the richer the model must be to accurately describe the behavior. Important questions are:

- Is the electric field constant or does it vary with time? At what rate?

- Does the response depend on the direction of the applied field (isotropy of the material)?

- Is the response the same everywhere (homogeneity of the material)?

- Do any boundaries or interfaces have to be taken into account?

- Is the response linear with respect to the field, or are there nonlinearities?

The relationship between the electric field E and the dipole moment M gives rise to the behavior of the dielectric, which, for a given material, can be characterized by the function F defined by the equation:

$$\mathbf{M} = \mathbf{F(E)}.$$

When both the type of electric field and the type of material have been defined, one then chooses the simplest function F that correctly predicts the phenomena of interest. Examples of phenomena that can be so modeled include:

- Refractive index

- Group velocity dispersion

- Birefringence

- Self-focusing

- Harmonic generation

Dipolar Polarization

Dipolar polarization is a polarization that is either inherent to polar molecules (orientation polarization), or can be induced in any molecule in which the asymmetric distortion of the nuclei is possible (distortion polarization). Orientation polarization results from a permanent dipole, e.g., that arising from the 104.45° angle between the asymmetric bonds between oxygen and hydrogen atoms in the water molecule, which retains polarization in the absence of an external electric field. The assembly of these dipoles forms a macroscopic polarization.

When an external electric field is applied, the distance between charges within each permanent dipole, which is related to chemical bonding, remains constant in orientation polarization; however, the direction of polarization itself rotates. This rotation occurs on a timescale that depends on the torque and surrounding local viscosity of the molecules. Because the rotation is not instantaneous, dipolar polarizations lose the response to electric fields at the highest frequencies. A molecule rotates about 1 radian per picosecond in a fluid, thus this loss occurs at about 10^{11} Hz (in the microwave region). The delay of the response to the change of the electric field causes friction and heat.

When an external electric field is applied at infrared frequencies or less, the molecules are bent and stretched by the field and the molecular dipole moment changes. The molecular vibration frequency is roughly the inverse of the time it takes for the molecules to bend, and this distortion polarization disappears above the infrared.

Ionic Polarization

Ionic polarization is polarization caused by relative displacements between positive and negative ions in ionic crystals (for example, NaCl).

If a crystal or molecule consists of atoms of more than one kind, the distribution of charges around an atom in the crystal or molecule leans to positive or negative. As a result, when lattice vibrations or molecular vibrations induce relative displacements of the atoms, the centers of positive and negative charges are also displaced. The locations of these centers are affected by the symmetry of the displacements. When the centers don't correspond, polarizations arise in molecules or crystals. This polarization is called ionic polarization.

Ionic polarization causes the ferroelectric effect as well as dipolar polarization. The ferroelectric transition, which is caused by the lining up of the orientations of permanent dipoles along a particular direction, is called an order-disorder phase transition. The transition caused by ionic polarizations in crystals is called a displacive phase transition.

Ionic Polarization of Cells

Ionic polarization enables the production of energy-rich compounds in cells (the proton pump in mitochondria) and, at the plasma membrane, the establishment of the resting potential, energetically unfavourable transport of ions, and cell-to-cell communication (the Na+/K+-ATPase).

All cells in animal body tissues are electrically polarized – in other words, they maintain a voltage difference across the cell's plasma membrane, known as the membrane potential. This electrical polarization results from a complex interplay between protein structures embedded in the membrane called ion pumps and ion channels.

In neurons, the types of ion channels in the membrane usually vary across different parts of the cell, giving the dendrites, axon, and cell body different electrical properties. As a result, some parts of the membrane of a neuron may be excitable (capable of generating action potentials), whereas others are not.

Dielectric Dispersion

In physics, dielectric dispersion is the dependence of the permittivity of a dielectric material on the frequency of an applied electric field. Because there is a lag between changes in polarization and changes in the electric field, the permittivity of the dielectric is a complicated function of frequency of the electric field. Dielectric dispersion is very important for the applications of dielectric materials and for the analysis of polarization systems.

This is one instance of a general phenomenon known as material dispersion: a frequency-dependent response of a medium for wave propagation.

When the frequency becomes higher:

1. dipolar polarization can no longer follow the oscillations of the electric field in the microwave region around 10^{10} Hz;

2. ionic polarization and molecular distortion polarization can no longer track the electric field past the infrared or far-infrared region around 10^{13} Hz, ;

3. electronic polarization loses its response in the ultraviolet region around 10^{15} Hz.

In the frequency region above ultraviolet, permittivity approaches the constant ε_o in every substance, where ε_o is the permittivity of the free space. Because permittivity indicates the strength of the relation between an electric field and polarization, if a polarization process loses its response, permittivity decreases.

Dielectric Relaxation

Dielectric relaxation is the momentary delay (or lag) in the dielectric constant of a material. This is usually caused by the delay in molecular polarization with respect to a changing electric field in a dielectric medium (e.g., inside capacitors or between two large conducting surfaces). Dielectric relaxation in changing electric fields could be considered analogous to hysteresis in changing magnetic fields (for inductors or transformers). Relaxation in general is a delay or lag in the response of a linear system, and therefore dielectric relaxation is measured relative to the expected linear steady state (equilibrium) dielectric values. The time lag between electrical field and polarization implies an irreversible degradation of Gibbs free energy.

In physics, dielectric relaxation refers to the relaxation response of a dielectric medium to an external, oscillating electric field. This relaxation is often described in terms of permittivity as a function of frequency, which can, for ideal systems, be described by the Debye equation. On the other hand, the distortion related to ionic and electronic polarization shows behavior of the resonance or oscillator type. The character of the distortion process depends on the structure, composition, and surroundings of the sample.

Debye Relaxation

Debye relaxation is the dielectric relaxation response of an ideal, noninteracting population of dipoles to an alternating external electric field. It is usually expressed in the complex permittivity ε of a medium as a function of the field's frequency ω:

$$\hat{\varepsilon}(\omega) = \varepsilon_\infty + \frac{\Delta\varepsilon}{1 + i\omega\tau},$$

where ε_∞ is the permittivity at the high frequency limit, $\Delta\varepsilon = \varepsilon_s - \varepsilon_\infty$ where ε_s is the static, low frequency permittivity, and τ is the characteristic relaxation time of the medium. Separating the real and imaginary parts of the complex dielectric permittivity yields:

$$\varepsilon' = \varepsilon_\infty + \frac{(\varepsilon_s - \varepsilon_\infty)}{(1 + \omega^2\tau^2)}$$

$$\varepsilon'' = \frac{(\varepsilon_s - \varepsilon_\infty)\omega\tau}{1 + \omega^2\tau^2}$$

The dielectric loss is also represented by:

$$\tan\delta = \frac{\varepsilon''}{\varepsilon'} = \frac{(\varepsilon_s - \varepsilon_\infty)\omega\tau}{\varepsilon_s + \varepsilon_\infty\omega^2\tau^2}$$

This relaxation model was introduced by and named after the physicist Peter Debye (1913). It is characteristic for dynamic polarization with only one relaxation time.

Variants of the Debye Equation

- Cole–Cole equation

This equation is used when the dielectric loss peak shows symmetric broadening.

- Cole–Davidson equation

This equation is used when the dielectric loss peak shows asymmetric broadening

- Havriliak–Negami relaxation

This equation considers both symmetric and asymmetric broadening

- Kohlrausch–Williams–Watts function (Fourier transform of stretched exponential function).

Paraelectricity

Paraelectricity is the ability of many materials (specifically ceramics) to become polarized under an applied electric field. Unlike ferroelectricity, this can happen even if there is no permanent electric dipole that exists in the material, and removal of the fields results in the polarization in the material returning to zero. The mechanisms that cause paraelectric behaviour are the distortion of individual ions (displacement of the electron cloud from the nucleus) and polarization of molecules or combinations of ions or defects.

Paraelectricity can occur in crystal phases where electric dipoles are unaligned and thus have the potential to align in an external electric field and weaken it.

An example of a paraelectric material of high dielectric constant is strontium titanate.

The $LiNbO_3$ crystal is ferroelectric below 1430 K, and above this temperature it transforms into a disordered paraelectric phase. Similarly, other perovskites also exhibit paraelectricity at high temperatures.

Paraelectricity has been explored as a possible refrigeration mechanism; polarizing a paraelectric by applying an electric field under adiabatic process conditions raises the temperature, while removing the field lowers the temperature. A heat pump that operates by polarizing the paraelectric, allowing it to return to ambient temperature (by dissipating the extra heat), bringing it into contact with the object to be cooled, and finally depolarizing it, would result in refrigeration.

Tunability

Tunable dielectrics are insulators whose ability to store electrical charge changes when a voltage is applied.

Generally, strontium titanate ($SrTiO_3$) is used for devices operating at low temperatures, while barium strontium titanate ($Ba_{1-x}Sr_xTiO_3$) substitutes for room temperature devices. Other potential materials include microwave dielectrics and carbon nanotube (CNT) composites.

In 2013 multi-sheet layers of strontium titanate interleaved with single layers of strontium oxide produced a dielectric capable of operating at up to 125 GHz. The material was created via molecular beam epitaxy. The two have mismatched crystal spacing that produces strain within the strontium titanate layer that makes it less stable and tunable.

Systems such as $Ba_{1-x}Sr_xTiO_3$ have a paraelectric–ferroelectric transition just below ambient temperature, providing high tunability. Such films suffer significant losses arising from defects.

Applications

Capacitors

Commercially manufactured capacitors typically use a solid dielectric material with high permittivity as the intervening medium between the stored positive and negative charges. This material is often referred to in technical contexts as the *capacitor dielectric*.

The most obvious advantage to using such a dielectric material is that it prevents the conducting plates, on which the charges are stored, from coming into direct electrical contact. More significantly, however, a high permittivity allows a greater stored charge at a given voltage. This can be seen by treating the case of a linear dielectric with permittivity ε and thickness d between two conducting plates with uniform charge density σ_ε. In this case the charge density is given by

$$\sigma_\varepsilon = \varepsilon \frac{V}{d}$$

and the capacitance per unit area by

$$c = \frac{\sigma_\varepsilon}{V} = \frac{\varepsilon}{d}$$

From this, it can easily be seen that a larger ε leads to greater charge stored and thus greater capacitance.

Dielectric materials used for capacitors are also chosen such that they are resistant to ionization. This allows the capacitor to operate at higher voltages before the insulating dielectric ionizes and begins to allow undesirable current.

Dielectric Resonator

A *dielectric resonator oscillator* (DRO) is an electronic component that exhibits resonance of the polarization response for a narrow range of frequencies, generally in the microwave band. It consists of a "puck" of ceramic that has a large dielectric constant and a low dissipation factor. Such resonators are often used to provide a frequency reference in an oscillator circuit. An unshielded dielectric resonator can be used as a dielectric resonator antenna (DRA).

Some Practical Dielectrics

Dielectric materials can be solids, liquids, or gases. In addition, a high vacuum can also be a useful, nearly lossless dielectric even though its relative dielectric constant is only unity.

Solid dielectrics are perhaps the most commonly used dielectrics in electrical engineering, and many solids are very good insulators. Some examples include porcelain, glass, and most plastics. Air, nitrogen and sulfur hexafluoride are the three most commonly used gaseous dielectrics.

- Industrial coatings such as parylene provide a dielectric barrier between the substrate and its environment.

- Mineral oil is used extensively inside electrical transformers as a fluid dielectric and to assist in cooling. Dielectric fluids with higher dielectric constants, such as electrical grade castor oil, are often used in high voltage capacitors to help prevent corona discharge and increase capacitance.

- Because dielectrics resist the flow of electricity, the surface of a dielectric may retain *stranded* excess electrical charges. This may occur accidentally when the dielectric is rubbed (the triboelectric effect). This can be useful, as in a Van de Graaff generator or electrophorus, or it can be potentially destructive as in the case of electrostatic discharge.

- Specially processed dielectrics, called electrets (which should not be confused with ferroelectrics), may retain excess internal charge or "frozen in" polarization. Electrets have a semipermanent electric field, and are the electrostatic equivalent to magnets. Electrets have numerous practical applications in the home and industry.

- Some dielectrics can generate a potential difference when subjected to mechanical stress, or (equivalently) change physical shape if an external voltage is applied across the material. This property is called piezoelectricity. Piezoelectric materials are another class of very useful dielectrics.

- Some ionic crystals and polymer dielectrics exhibit a spontaneous dipole moment, which can be reversed by an externally applied electric field. This behavior is called the ferroelectric effect. These materials are analogous to the way ferromagnetic materials behave within an externally applied magnetic field. Ferroelectric materials often have very high dielectric constants, making them quite useful for capacitors.

Gaseous Dielectrics

- Atmospheric air is the cheapest and most widely used dielectric . Other gaseous dielectrics, used as compressed gas at higher pressures than atmospheric in power system, are Nitrogen, Sulphurhexafluoride SF_6 (an electro-negative gas), and

- It's mixtures with CO_2 and N_2. SF_6 is very widely applied for Gas Insulated Systems (GIS), Circuit Breakers and gas filled installations i.e. sub-stations and cables. It is being now applied for power transformers also.

Dielectric Gas

A dielectric gas, or insulating gas, is a dielectric material in gaseous state. Its main purpose is to prevent or rapidly quench electric discharges. Dielectric gases are used as electrical insulators in high voltage applications, e.g. transformers, circuit breakers (namely sulfur hexafluoride circuit breakers), switchgear (namely high voltage switchgear), radar waveguides, etc.

A good dielectric gas should have high dielectric strength, high thermal stability and chemical inertness against the construction materials used, non-flammability and low toxicity, low boiling point, good heat transfer properties, and low cost.

The most common dielectric gas is air, due to its ubiquity and low cost. Another commonly used gas is a dry nitrogen.

In special cases, e.g., high voltage switches, gases with good dielectric properties and very high breakdown voltages are needed. Highly electronegative elements, e.g., halogens, are favored as they rapidly recombine with the ions present in the discharge channel. The halogen gases are highly corrosive. Other compounds, which dissociate only in the discharge pathway, are therefore preferred; sulfur hexafluoride, organofluorides (especially perfluorocarbons) and chlorofluorocarbons are the most common.

The breakdown voltage of gases is roughly proportional to their density. Breakdown voltages also increase with the gas pressure; many gases however have limited upper pressure due to their liquefaction.

The decomposition products of halogenated compounds are highly corrosive, the occurrence of corona discharge should therefore be prevented.

Build-up of moisture can degrade dielectric properties of the gas. Moisture analysis is used for early detection of this.

Dielectric gases can also serve as coolants.

Vacuum is an alternative for gas in some applications.

Mixtures of gases can be used where appropriate. Addition of sulfur hexafluoride can dramatically improve the dielectric properties of poorer insulators, e.g. helium or nitrogen. Multicomponent gas mixtures can offer superior dielectric properties; the optimum mixtures combine the electron attaching gases (sulfur hexafluoride, octafluorocyclobutane) with molecules capable of thermalizing (slowing down) accelerated electrons (e.g. tetrafluoromethane, fluoroform. The insulator properties of the gas are controlled by the combination of electron attachment, electron scattering, and electron ionization.

Atmospheric pressure significantly influences the insulation properties of air. High-voltage applications, e.g. xenon flash lamps, can experience electrical breakdowns at high altitudes.

Electrical Insulation and Dielectrics

Vacuum as Dielectric

- Vacuum of the order of 10^{-5} Torr and lower provides an excellent electrical insulation. Vacuum technology developed and applied for circuit breakers in the last three decades is phenomenon.

Utilization of Dielectric Properties

The value of η also represents the degree of utilization of the dielectric in between two elecrtodes. A higher value of η represents better utilization of the insulating properties of a dielectric. It compares the ideal condition of electric field intensity (uniform field between electrodes at the same distance d apart) with the existing actual maximum field intensity. Thus η, a dimensionless quantity enables a comparison of the degree of uniformity of field configurations formed between different electrodes. The value of η lies between, $0 \leq \eta \leq 1$.

With the knowledge of the value of η for a particular field configuration, the maximum electric field intensity or the maximum electric stress on a dielectric can easily be estimated. η serves as a ready reference which is an important information for insulation design in equipment. However, for determining the exact magnitude of maximum electric stress, at different shapes of electrodes used in the equipment, numerical estimation techniques have to be applied.

Stress Control

- More the uniformity in field, better is the utilisation of the dielectric.

- An ideal utilisation is accomplished only where η is equal to one, which is not possible in practice.

- More nonuniform field represents higher electric stress in the dielectric. It could be at only a particular location. Insulation design in an equipment is made with due consideration to the value of estimated maximum electric field intensity.

- It is possible to achieve a higher degree of uniformity of fields by giving suitable shapes and sizes to various electrodes in an equipment.

    For example, abrupt interruption of electrodes, both anode or cathode, in high voltage equipment leads to concentration of electric field at the brim, resulting in a tremendous enhancement of electric stress on the dielectric. The dielectric in the vicinity thus becomes highly vulnerable to breakdown.

- The electrodes must be given a suitable shape at the brim to control the stress.

- For stress control, in principle the electrodes are extended and formed in such a way that higher field intensity than in the main field region does not appear anywhere in the dielectric.

- Rogowski suggested in 1923, a shape by which the electrodes could be extended, known as 'Rogowski Profile'.

- One can see in this figure that the field intensity continuously reduces beyond the main field region.

- Another shape of the electrode credited to Borda known as 'Borda Profile' , figure on the left, was actually worked out by him in as early as 1766 in France, more than 200 years ago.

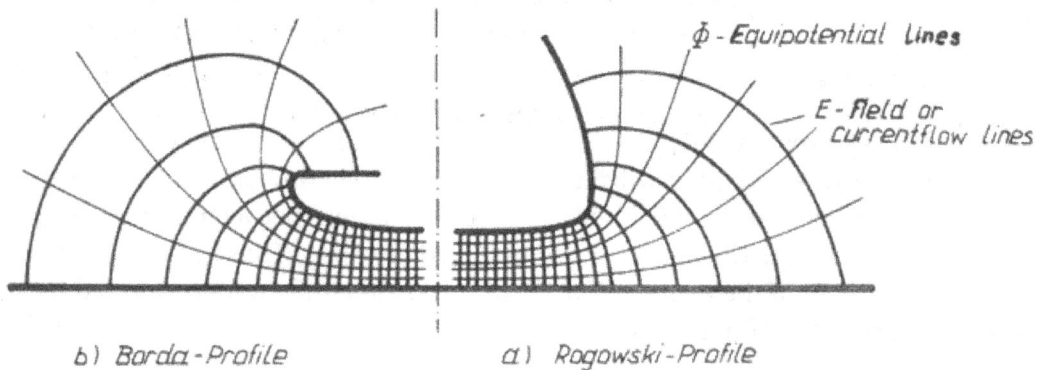

Fig. i Equipotential and Field (current flow) lines between plane and brim field

- Electrodes at high potentials in the laboratory are given large, smooth shaped dome like bodies or shapes like toroids to bring down electric stress on the atmospheric air (dielectric). The modern trend in such electrode design includes 'segmented electrodes', constituting a number of small, identical, smooth discs given a large desired continuous shape as per requirement. The curvatures of the individual segment discs are worked out by optimisation of the suggested profiles. Figure ii shows both, single metallic body and segmented electrodes used in HV test apparatus for stress control.

- Extended shapes of electrodes, also known as 'shields', are suitably provided on high voltage apparatus for electric stress control as shown in the figure iii. Sharp contacts are often enveloped by a large diameter hemispherical electrode having an aperture, or provided with concentric toroidal rings (doughnut shaped ring). Spheres with smooth holes are provided at bends for the connections of circular and tubular electrodes. Instead of wires, tubular electrodes of large diameters are used for connections in high voltage laboratories which bring down the field intensity at higher voltages considerably. These measures are necessary not only to prevent any partial breakdown (corona) occurring in the laboratory but also to check radio interference.

- It is a common practice to use bundles of two or more number of conductors at the same potential instead of a single conductor, to bring down the electric stress, i.e., for stress control, (Fig. v, b). As the transmission voltages are increasing, bundles with eight or even more number of conductors are being used at higher voltages.

- Capacitive grading is provided in high voltage bushings, potential transformers and cable terminations in order to achieve a better potential distribution leading to a more uniform field distribution in the dielectric. It is achieved by inserting concentric conductive layers at appropriate positions, known as 'floating screens', to control the electric stress as shown in Fig. iv. This enables an economic utilisation of the insulating material by evenly distributing the equipotential surfaces in the complete dielectric.

- Use of screen (also known as concentric conductor at ground potential) over the insulation in coaxial high voltage cables is made to control the electric stress. The field achieved in these screened cables is radial and generally a weakly nonuniform field. For making cable joints and terminations, the,screen is extended in the form of a cone, known as 'stress cone'. This helps in achieving a more uniform distribution of electric stress in the dielectric at cable end termination as shown in Fig. iii (d).

- A modest thumb rule to control electric stress in high voltage apparatus is to avoid sharp points and edges. Symmetrical, smooth shaped and large electrodes are preferable. It must be borne in mind that even the roughness on metallic surfaces can lead to distortions in the field at higher voltages. Furthermore, microprotrusions may grow and penetrate deeper in the dielectric leading to excessive field enhancement. These must be prevented from developing, firstly during manufacturing stage and subsequently during service and maintenance.

Segmented electrodes(Complete HV lab (600 kV AC))   Single metallic body (DC Generator 900 kV)

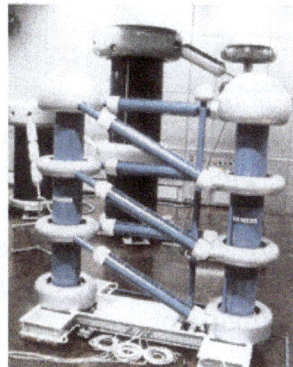

fig. ii

(a) Transformer Bushing   (b) GIS

(c) HV Capacitor   (d) Cable Termination

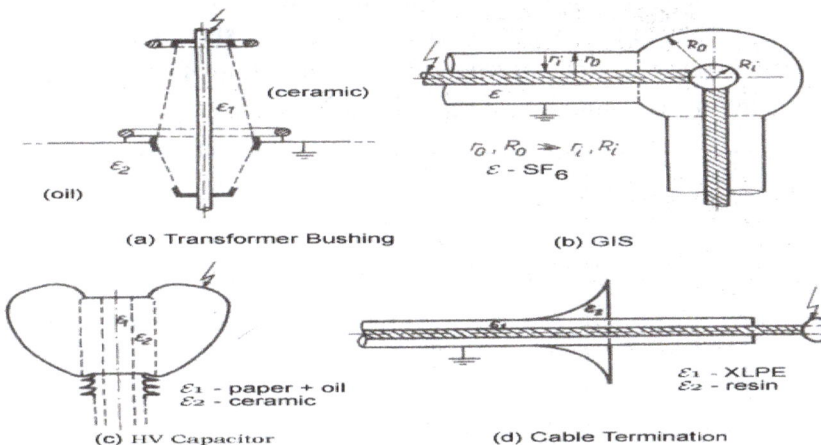

Fig. iii Extended shapes of electrodes for stress control (a) A bushing with toroids (b) Right angle bend of a bus bar in gas insulated switchgear (GIS), (c) HV electrode on a condenser, (d) stress cone at a screened cable end.

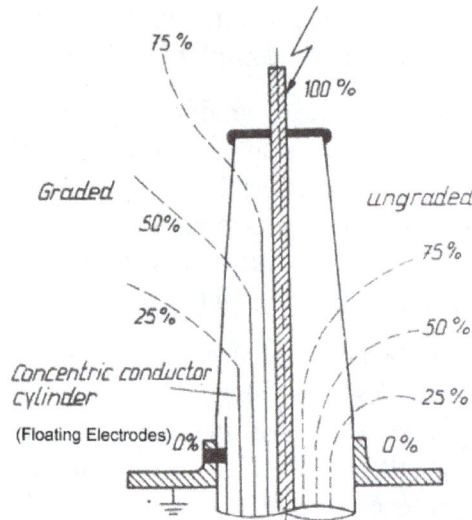

Fig. iv Potential distribution in a bushing with and without capacitive grading

Ferroelectricity

Ferroelectricity is a property of certain materials that have a spontaneous electric polarization that can be reversed by the application of an external electric field. The term is used in analogy to ferromagnetism, in which a material exhibits a permanent magnetic moment. Ferromagnetism was already known when ferroelectricity was discovered in 1920 in Rochelle salt by Valasek. Thus, the prefix *ferro*, meaning iron, was used to describe the property despite the fact that most ferroelectric materials do not contain iron.

Polarization

When most materials are polarized, the polarization induced, P, is almost exactly proportional to the applied external electric field E; so the polarization is a linear function. This is called dielectric polarization. Some materials, known as paraelectric materials, show a more enhanced nonlinear polarization. The electric permittivity, corresponding to the slope of the polarization curve, is not constant as in dielectrics but is a function of the external electric field.

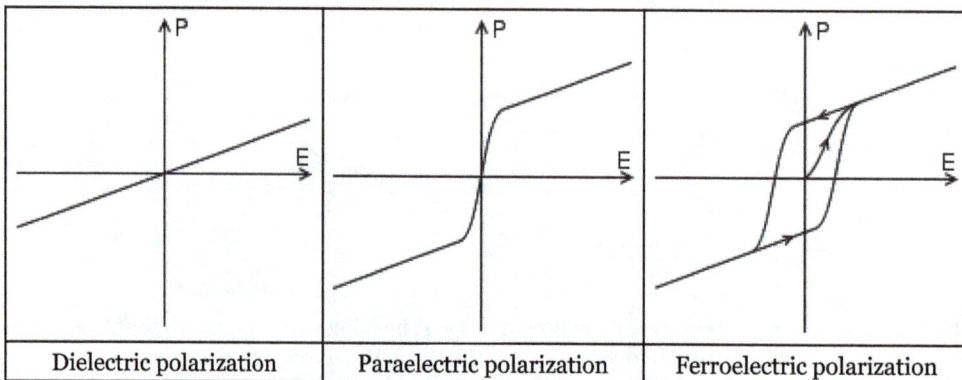

| Dielectric polarization | Paraelectric polarization | Ferroelectric polarization |

In addition to being nonlinear, ferroelectric materials demonstrate a spontaneous nonzero polarization even when the applied field E is zero. The distinguishing feature of ferroelectrics is that the spontaneous polarization can be *reversed* by a suitably strong applied electric field in the opposite direction; the polarization is therefore dependent not only on the current electric field but also on its history, yielding a hysteresis loop. They are called ferroelectrics by analogy to ferromagnetic materials, which have spontaneous magnetization and exhibit similar hysteresis loops.

Typically, materials demonstrate ferroelectricity only below a certain phase transition temperature, called the Curie temperature, T_c, and are paraelectric above this temperature: the spontaneous polarization vanishes, and the ferroelectric crystal transforms into the paraelectric state. Many ferroelectrics lose their piezoelectric properties above Tc completely, because their paraelectric phase has centrosymmetric crystallographic structure.

Applications

The nonlinear nature of ferroelectric materials can be used to make capacitors with tunable capacitance. Typically, a ferroelectric capacitor simply consists of a pair of electrodes sandwiching a layer of ferroelectric material. The permittivity of ferroelectrics is not only tunable but commonly also very high in absolute value, especially when close to the phase transition temperature. Because of this, ferroelectric capacitors are small in physical size compared to dielectric (non-tunable) capacitors of similar capacitance.

The spontaneous polarization of ferroelectric materials implies a hysteresis effect which can be used as a memory function, and ferroelectric capacitors are indeed used to make ferroelectric RAM for computers and RFID cards. In these applications thin films of ferroelectric materials are typically used, as this allows the field required to switch the polarization to be achieved with a moderate voltage. However, when using thin films a great deal of attention needs to be paid to the interfaces, electrodes and sample quality for devices to work reliably.

Ferroelectric materials are required by symmetry considerations to be also piezoelectric and pyroelectric. The combined properties of memory, piezoelectricity, and pyroelectricity make ferroelectric capacitors very useful, e.g. for sensor applications. Ferroelectric capacitors are used in medical ultrasound machines (the capacitors generate and then listen for the ultrasound ping used to image the internal organs of a body), high quality infrared cameras (the infrared image is projected onto a two dimensional array of ferroelectric capacitors capable of detecting temperature differences as small as millionths of a degree Celsius), fire sensors, sonar, vibration sensors, and even fuel injectors on diesel engines.

Another idea of recent interest is the *ferroelectric tunnel junction (FTJ)* in which a contact made up by nanometer-thick ferroelectric film placed between metal electrodes. The thickness of the ferroelectric layer is small enough to allow tunneling of electrons. The piezoelectric and interface effects as well as the depolarization field may lead to a giant electroresistance (GER) switching effect.

Yet another hot topic is multiferroics, where researchers are looking for ways to couple magnetic and ferroelectric ordering within a material or heterostructure; there are several recent reviews on this topic.

Materials

The internal electric dipoles of a ferroelectric material are coupled to the material lattice so anything that changes the lattice will change the strength of the dipoles (in other words, a change in the spontaneous polarization). The change in the spontaneous polarization results in a change in the surface charge. This can cause current flow in the case of a ferroelectric capacitor even without the presence of an external voltage across the capacitor. Two stimuli that will change the lattice dimensions of a material are force and temperature. The generation of a surface charge in response to the application of an external stress to a material is called piezoelectricity. A change in the spontaneous polarization of a material in response to a change in temperature is called pyroelectricity.

Generally, there are 230 space groups among which 32 crystalline classes can be found in crystals. There are 21 non-centrosymmetric classes, within which 20 are piezoelectric. Among the piezoelectric classes, 10 have a spontaneous electric polarization, that varies with the temperature, therefore they are pyroelectric. Among pyroelectric materials, some of them are ferroelectric.

32 Crystalline classes				
21 noncentrosymmetric				11 centrosymmetric
20 classes piezoelectric				
10 classes pyroelectric		non pyroelectric		non piezoelectric
ferroelectric	non ferroelectric			
e.g. : $PbZr/TiO_3$, Ba-TiO_3, $PbTiO_3$	e.g. : Tourmaline, ZnO, AlN	e.g. : Quartz, Langasite		

Ferroelectric phase transitions are often characterized as either displacive (such as $BaTiO_3$) or order-disorder (such as $NaNO_2$), though often phase transitions will demonstrate elements of both behaviors. In barium titanate, a typical ferroelectric of the displacive type, the transition can be understood in terms of a polarization catastrophe, in which, if an ion is displaced from equilibrium slightly, the force from the local electric fields due to the ions in the crystal increases faster than the elastic-restoring forces. This leads to an asymmetrical shift in the equilibrium ion positions and hence to a permanent dipole moment. The ionic displacement in barium titanate concerns the relative position of the titanium ion within the oxygen octahedral cage. In lead titanate, another key ferroelectric material, although the structure is rather similar to barium titanate the driving force for ferroelectricity is more complex with interactions between the lead and oxygen ions also playing an important role. In an order-disorder ferroelectric, there is a dipole moment in each unit cell, but at high temperatures they are pointing in random directions. Upon lowering the temperature and going through the phase transition, the dipoles order, all pointing in the same direction within a domain.

An important ferroelectric material for applications is lead zirconate titanate (PZT), which is part of the solid solution formed between ferroelectric lead titanate and anti-ferroelectric lead zirconate. Different compositions are used for different applications; for memory applications, PZT closer in composition to lead titanate is preferred, whereas piezoelectric applications make use of the diverging piezoelectric coefficients associated with the morphotropic phase boundary that is found close to the 50/50 composition.

Ferroelectric crystals often show several transition temperatures and domain structure hysteresis, much as do ferromagnetic crystals. The nature of the phase transition in some ferroelectric crystals is still not well understood.

In 1974 R.B. Meyer used symmetry arguments to predict ferroelectric liquid crystals, and the prediction could immediately be verified by several observations of behavior connected to ferroelectricity in smectic liquid-crystal phases that are chiral and tilted. The technology allows the building of flat-screen monitors. Mass production between 1994 and 1999 was carried out by Canon. Ferroelectric liquid crystals are used in production of reflective LCoS.

In 2010 David Field found that prosaic films of chemicals such as nitrous oxide or propane exhibited ferroelectric properties. This new class of ferroelectric materials exhibit "spontelectric" properties, and may have wide ranging applications in device and nano-technology and also influence the electrical nature of dust in the interstellar medium.

Other ferroelectric materials used include triglycine sulfate, polyvinylidene fluoride (PVDF) and lithium tantalate.

Theory

An introduction to Landau theory can be found here. Based on Ginzburg–Landau theory, the free energy of a ferroelectric material, in the absence of an electric field and applied stress may be written as a Taylor expansion in terms of the order parameter, P. If a sixth order expansion is used (i.e. 8th order and higher terms truncated), the free energy is given by:

$$\Delta E = \frac{1}{2}\alpha_0\left(T-T_0\right)\left(P_x^2 + P_y^2 + P_z^2\right) + \frac{1}{4}\alpha_{11}\left(P_x^4 + P_y^4 + P_z^4\right)$$

$$+ \frac{1}{2}\alpha_{12}\left(P_x^2 P_y^2 + P_y^2 P_z^2 + P_z^2 P_x^2\right)$$

$$+ \frac{1}{6}\alpha_{111}\left(P_x^6 + P_y^6 + P_z^6\right)$$

$$+ \frac{1}{2}\alpha_{112}\left[P_x^4\left(P_y^2 + P_z^2\right) + P_y^4\left(P_x^2 + P_z^2\right) + P_z^4\left(P_x^2 + P_y^2\right)\right]$$

$$+ \frac{1}{2}\alpha_{123}P_x^2 P_y^2 P_z^2$$

where P_x, P_y, and P_z are the components of the polarization vector in the x, y, and z directions respectively, and the coefficients, $\alpha_i, \alpha_{ij}, \alpha_{ijk}$ must be consistent with the crystal symmetry. To investigate domain formation and other phenomena in ferroelectrics, these equations are often used in the context of a phase field model. Typically, this involves adding a gradient term, an electrostatic term and an elastic term to the free energy. The equations are then discretized onto a grid using the finite difference method and solved subject to the constraints of Gauss's law and Linear elasticity.

In all known ferroelectrics, $\alpha_0 > 0$ and $\alpha_{111} > 0$. These coefficients may be obtained experimentally or from ab-initio simulations. For ferroelectrics with a first order phase transition, $\alpha_{11} < 0$, whereas $\alpha_{11} > 0$ for a second order phase transition.

The spontaneous polarization, P_s of a ferroelectric for a cubic to tetragonal phase transition may be obtained by considering the 1D expression of the free energy which is:

$$\Delta E = \frac{1}{2}\alpha_0 \left(T-T_0\right)P_x^2 + \frac{1}{4}\alpha_{11}P_x^4 + \frac{1}{6}\alpha_{111}P_x^6$$

This free energy has the shape of a double well potential with two free energy minima at $P = \pm P_s$, where P_s is the spontaneous polarization. At these two minima, the derivative of the free energy is zero, i.e.:

$$\frac{\partial \Delta E}{\partial P_x} = \alpha_0 \left(T-T_0\right)P_x + \alpha_{11}P_x^3 + \alpha_{111}P_x^5 = 0$$

$$P_x\left[\alpha_0 \left(T-T_0\right) + \alpha_{11}P_x^2 + \alpha_{111}P_x^4\right] = 0$$

Since $P_x = 0$ corresponds to a free energy maxima in the ferroelectric phase, the spontaneous polarization, P_s, is obtained from the solution of the equation:

$$\alpha_0 \left(T-T_0\right) + \alpha_{11}P_x^2 + \alpha_{111}P_x^4 = 0$$

which is:

$$P_s^2 = \frac{1}{2\alpha_{111}}\left[-\alpha_{11} \pm \sqrt{\alpha_{11}^2 - 4\alpha_0\alpha_{111}\left(T-T_0\right)}\right]$$

and elimination of solutions yielding a negative square root (for either the first or second order phase transitions) gives:

$$P_s = \sqrt{\frac{1}{2\alpha_{111}}\left[-\alpha_{11} + \sqrt{\alpha_{11}^2 - 4\alpha_0\alpha_{111}\left(T-T_0\right)}\right]}$$

If $\alpha_{11} = 0$, using the same approach as above, the spontaneous polarization may be obtained as:

$$P_s = \sqrt{-\frac{\alpha_0 \left(T-T_0\right)}{\alpha_{111}}}$$

The hysteresis loop (P_x versus E_x) may be obtained from the free energy expansion by adding another electrostatic term, $E_x P_x$, as follows:

$$\Delta E = \frac{1}{2}\alpha_0 \left(T-T_0\right)P_x^2 + \frac{1}{4}\alpha_{11}P_x^4 + \frac{1}{6}\alpha_{111}P_x^6 - E_x P_x$$

$$\frac{\partial \Delta E}{\partial P_x} = \alpha_0 \left(T-T_0\right)P_x + \alpha_{11}P_x^3 + \alpha_{111}P_x^5 - E_x = 0$$

$$E_x = \alpha_0 \left(T - T_0 \right) P_x + \alpha_{11} P_x^3 + \alpha_{111} P_x^5$$

Plotting E_x as a function of P_x and reflecting the graph about the 45 degree line gives an 'S' shaped curve. The central part of the 'S' corresponds to a free energy local maximum (since $\dfrac{\partial^2 \Delta E}{\partial P_x^2} < 0$). Elimination of this region, and connection of the top and bottom portions of the 'S' curve by vertical lines at the discontinuities gives the hysteresis loop.

Faraday described the space around a magnet to be filled with 'lines of magnetic force'. Similarly, the region around an electrified object may be considered to be filled with 'lines of electric force'. To Faraday, these lines existed as mechanical structures in the surrounding medium (the dielectric) and could exert force on an object placed therein. Two typical electrostatic field structures are shown in Fig. V. In figure (a), the field between a sphere or a cylinder and plane is sketched, and in figure(b), the field pattern on a bundle of four conductors used for transmission lines is shown, neglecting the effect of ground.

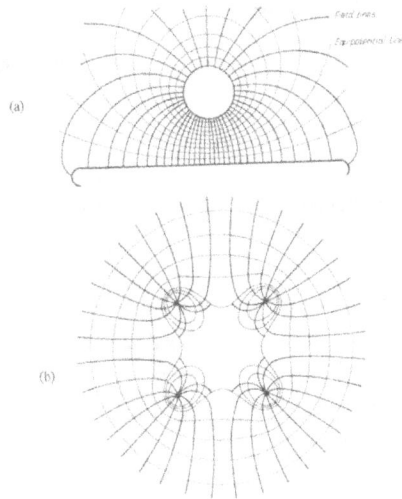

Fig. v Typical electrostatic field configurations. (a) Field between sphere or cylinder and plane, (b) Field on a bundle of four conductors

- The 'electric field intensity', also known as the' electric field strength', is defined as the electrostatic force F exerted by the field on a unit positive test charge q, placed at a particular point P in a dielectric. It is denoted by E, and expressed in unit 'Newtons per Coulomb', that is, the force per unit charge.

- The electric field intensity is measured in its practical units of 'Volts per meter' (V/m or kV/mm).

- The electric field intensity is often more specifically mentioned as 'electric stress' experienced by a dielectric or an electrical insulating material.

- The potential difference between two points a and b, having scalar potential ϕ_a, ϕ_b in a space charge free electric field $\overline{E}$, is defined as the work done by an external source in moving a unit positive charge from b to a,

$$U_{ab} = -\int_b^a |\overline{E}| dx = (\phi_a - \phi_b)$$

- U_{ab} is positive if the work is done in carrying the positive charge from b to a. The maximum magnitude of electric field intensity is therefore, given by the maximum value of the rate of change of potential with distance. It is obtained when the direction of the increment of distance is opposite to the direction of $\overline{E}$, in other words, the maximum value of the rate of change of potential is obtained when the direction of $\overline{E}$ is opposite to the direction in which the potential is increasing most rapidly,

$$\frac{dU_{ab}}{dx}\Big|_{max} = -|\overline{E}|_{max}$$

The operator on ϕ by which $\overline{E}$ is obtained, is thus known as the 'gradient'. The relationship between ϕ and $\overline{E}$ may be written as,

$$\overline{E} = -\text{grad}\phi$$

- The electric field intensity is, therefore, numerically equal to the 'potential gradient'.

Electric Strength of Dielectrics

- The qualitative definition of 'electric strength' of a dielectric is 'the maximum electric stress a dielectric can withstand'.

- A large number of factors affect the electric breakdown of a dielectric, these include pressure, humidity, temperature, electric field configuration (electrode shape and size) electrode material, applied voltage waveform, its duration and magnitude, presence of impurities and imperfections in the dielectric, the composition of dielectric material. Hence a quantitative definition is complicated.

- In a time varying ac power frequency field (quasi stationary field), the maximum electric stress occur at the peak value of the applied voltage.

- Intrinsic strength of a dielectric: It is defined for gaseous and other than gaseous dielectric differently.

  o Gaseous dielectric: It is the magnitude of breakdown voltage measured across a gap distance of one cm in uniform field (η = 1) at normal temperature and pressure.

  o Liquid and Solid dielectrics: It is the highest value of breakdown strength obtained after eliminating all known secondary effects which may influence the breakdown adversely.

  It is measured for the ideal conditions of the dielectric in uniform field. Since it is very very high for solid and liquid dielectrics compared to gaseous dielectrics, it is measured for mm and μm thin films of the liquid and solid dielectrics respectively instead of 1 cm gap distance in case of gaseous dielectrics.

Cole–Cole Equation

The Cole–Cole equation is a relaxation model that is often used to describe dielectric relaxation in polymers.

It is given by the equation

$$\varepsilon^*(\omega) - \varepsilon_\infty = \frac{\varepsilon_s - \varepsilon_\infty}{1 + (i\omega\tau)^{1-\alpha}}$$

where ε^* is the complex dielectric constant, ε_s and ε_∞ are the "static" and "infinite frequency" dielectric constants, ω is the angular frequency and τ is a time constant.

The exponent parameter α, which takes a value between 0 and 1, allows to describe different spectral shapes. When $\alpha = 0$, the Cole-Cole model reduces to the Debye model. When $\alpha > 0$, the relaxation is *stretched*, i.e. it extends over a wider range on a logarithmic ω scale than Debye relaxation.

The separation of the complex dielectric constant $\varepsilon(\omega)$ was reported in the original paper by Cole and Cole as follows:

$$\varepsilon' = \varepsilon_\infty + (\varepsilon_s - \varepsilon_\infty) \frac{1 + (\omega\tau)^{1-\alpha} \sin \alpha\pi/2}{1 + 2(\omega\tau)^{1-\alpha} \sin \alpha\pi/2 + (\omega\tau)^{2(1-\alpha)}}$$

$$\varepsilon'' = \frac{(\varepsilon_s - \varepsilon_\infty)(\omega\tau)^{1-\alpha} \cos \alpha\pi/2}{1 + 2(\omega\tau)^{1-\alpha} \sin \alpha\pi/2 + (\omega\tau)^{2(1-\alpha)}}$$

Upon introduction of hyperbolic functions, the above expressions reduce to:

$$\varepsilon' - \varepsilon_\infty = \frac{1}{2}(\varepsilon_0 - \varepsilon_\infty) \left[1 - \frac{\sinh(1-\alpha)x}{\cosh(1-\alpha)x + \cos \alpha\pi/2} \right]$$

$$\varepsilon'' = \frac{\frac{1}{2}(\varepsilon_0 - \varepsilon_\infty) \cos \alpha\pi/2}{\cosh(1-\alpha)x + \sin \alpha\pi/2}$$

Here $x = \ln(\omega\tau)$.

These equations reduce to the Debye expression when $\alpha = 0$.

Cole-Cole relaxation constitutes a special case of Havriliak-Negami relaxation when the symmetry parameter (β) is equal to 1 - that is, when the relaxation peaks are symmetric. Another special case of Havriliak-Negami relaxation ($\beta<1$, $\alpha=1$) is known as Cole-Davidson relaxation, for an abridged and updated review of anomalous dielectric relaxation in dissored systems.

Havriliak–Negami Relaxation

Havriliak–Negami relaxation is an empirical modification of the Debye relaxation model, accounting for the asymmetry and broadness of the dielectric dispersion curve. The model was first used to describe the dielectric relaxation of some polymers, by adding two exponential parameters to the Debye equation:

$$\hat{\varepsilon}(\omega) = \varepsilon_\infty + \frac{\Delta\varepsilon}{(1+(i\omega\tau)^\alpha)^\beta},$$

where ε_∞ is the permittivity at the high frequency limit, $\Delta\varepsilon = \varepsilon_s - \varepsilon_\infty$ where ε_s is the static, low frequency permittivity, and τ is the characteristic relaxation time of the medium. The exponents α and β describe the asymmetry and broadness of the corresponding spectra.

Depending on application, the Fourier transform of the stretched exponential function can be a viable alternative that has one parameter less.

For $\beta = 1$ the Havriliak–Negami equation reduces to the Cole–Cole equation, for $\alpha = 1$ to the Cole–Davidson equation.

Mathematical Properties

Real and Imaginary Parts

The storage part ε' and the loss part ε'' of the permittivity (here: $\hat{\varepsilon}(\omega) = \varepsilon'(\omega) - i\varepsilon''(\omega)$) can be calculated as

$$\varepsilon'(\omega) = \varepsilon_\infty + \Delta\varepsilon\left(1 + 2(\omega\tau)^\alpha \cos(\pi\alpha/2) + (\omega\tau)^{2\alpha}\right)^{-\beta/2} \cos(\beta\phi)$$

and

$$\varepsilon''(\omega) = \Delta\varepsilon\left(1 + 2(\omega\tau)^\alpha \cos(\pi\alpha/2) + (\omega\tau)^{2\alpha}\right)^{-\beta/2} \sin(\beta\phi)$$

with

$$\phi = \arctan\left(\frac{(\omega\tau)^\alpha \sin(\pi\alpha/2)}{1 + (\omega\tau)^\alpha \cos(\pi\alpha/2)}\right)$$

Loss Peak

The maximum of the loss part lies at

$$\omega_{max} = \left(\frac{\sin\left(\dfrac{\pi\alpha}{2(\beta+1)}\right)}{\sin\left(\dfrac{\pi\alpha\beta}{2(\beta+1)}\right)}\right)^{1/\alpha} \tau^{-1}$$

Superposition of Lorentzians

The Havriliak–Negami relaxation can be expressed as a superposition of individual Debye relaxations

$$\frac{\hat{\varepsilon}(\omega)-\epsilon_\infty}{\Delta\varepsilon} = \int_{\tau_D=0}^{\infty} \frac{1}{1+i\omega\tau_D} g(\ln\tau_D)d\ln\tau_D$$

with the distribution function

$$g(\ln\tau_D) = \frac{1}{\pi} \frac{(\tau_D/\tau)^{\alpha\beta}\sin(\beta\theta)}{((\tau_D/\tau)^{2\alpha}+2(\tau_D/\tau)^\alpha\cos(\pi\alpha)+1)^{\beta/2}}$$

where

$$\theta = \arctan\left(\frac{\sin(\pi\alpha)}{(\tau_D/\tau)^\alpha+\cos(\pi\alpha)}\right)$$

if the argument of the arctangent is positive, else

$$\theta = \arctan\left(\frac{\sin(\pi\alpha)}{(\tau_D/\tau)^\alpha+\cos(\pi\alpha)}\right)+\pi$$

Logarithmic Moments

The first logarithmic moment of this distribution, the average logarithmic relaxation time is

$$\langle\ln\tau_D\rangle = \ln\tau + \frac{\Psi(\beta)+Eu}{\alpha}$$

where Ψ is the digamma function and Eu the Euler constant.

Inverse Fourier Transform

The inverse Fourier transform of the Havriliak-Negami function (the corresponding time-domain relaxation function) can be numerically calculated. It can be shown that the series expansions involved are special cases of the Fox-Wright function. In particular, in the time-domain the corresponding of $\hat{\varepsilon}(\omega)$ can be represented as

$$X(t) = \varepsilon_\infty\delta(t) + \frac{\Delta\varepsilon}{\tau}\left(\frac{t}{\tau}\right)^{\alpha\beta-1} E^\beta_{\alpha,\alpha\beta}(-(t/\tau)^\alpha),$$

where $\delta(t)$ is the Dirac delta function and

$$E^\gamma_{\alpha,\beta}(z) = \frac{1}{\Gamma(\gamma)}\sum_{k=0}^{\infty}\frac{\Gamma(\gamma+k)z^k}{k!\Gamma(\alpha k+\beta)}$$

is a special instance of the Fox-Wright function and, precisely, it is the three parameters Mittag-Leffler function also known as the Prabhakar function. The function $E_{\alpha,\beta}^{\gamma}(z)$ can be numerically evaluated, for instance, by means of a Matlab code .

Insulation Resistance Offered by Dielectrics

The concept of insulation resistance of a dielectric is represented by the dc resistance offered by an insulating material. It is generally described as specific insulation resistance 'ρ_{ins}', which is reciprocal of the dc conductivity κ_{dc},

$$\rho_{ins} = \frac{1}{K_{dc}} \qquad \Omega.m$$

Consider a direct voltage U_{dc} applied across two uniform field electrodes separated by a block of insulating material having an area A and length d as shown in the figure below. From the equivalent circuit diagram constituting a capacitance C and a dc resistance R_{dc} in parallel, following relation can be derived:

$$R_{dc} = \rho_{ins}\frac{d}{A}$$

$$i_{dc} = \frac{U}{R_{dc}} = \frac{U.A}{\rho_{ins}.d}$$

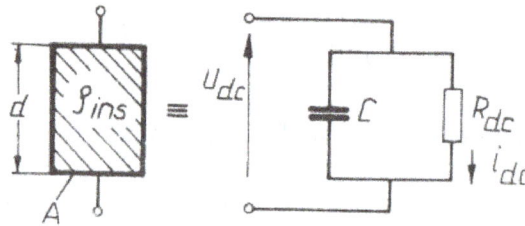

An insulating material and its equivalent circuit diagram

for a uniform field where

$$E = \frac{U}{d}$$

$$i_{dc} = \frac{E.A.d}{\rho_{ins}.d} = \frac{E.A}{\rho_{ins}}$$

$$= K_{dc}.A.E$$

Like κ_{dc}, the specific insulation resistance also depends strongly upon the temperature as shown in figure for transformer oil.

The specific insulation resistance, or the dc conductivity is a function of time for which the voltage is applied. A schematic illustration of the variation of κ_{dc} with respect to the time of application of the direct voltage for an insulating oil is shown in the following figure. The dc conductivity, which is very high initially (region A), is determined by the orientation of dipoles.

In region B, the conductivity is determined by the movement of free charge carriers under the influence of applied electric field. The magnitude of conductivity in this region also represents the ac power frequency conductivity 'κ_{ac}'. The region C in the figure represents the development of space charge in front of the electrodes. The steady ion current due to dissociation is depicted by region D.

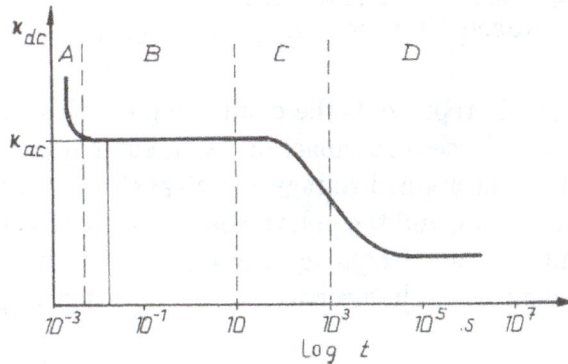

A schematic of dc conductivity of an insulating oil with respect to the time of applied voltage

In every dielectric matter, including those which have a low concentration of free charge carriers, a conductive current through the dielectric is always present on applying a voltage across the volume of the dielectric. Besides, small currents may also flow along the surfaces of the dielectric. Accordingly, two different conductivities, the volume and the surface conductivities of a dielectric, are distinguished. Reciprocal of these are the specific volume and surface insulation resistance's, respectively. The specific insulation resistance ρ insdescribed above represents the specific volume resistance 'ρ_v' of the dielectric having a unit of $\Omega.m$. The specific surface resistance 'ρ_s' has accordingly a unit of Ω only.

Dielectric Power Losses in Insulating Materials

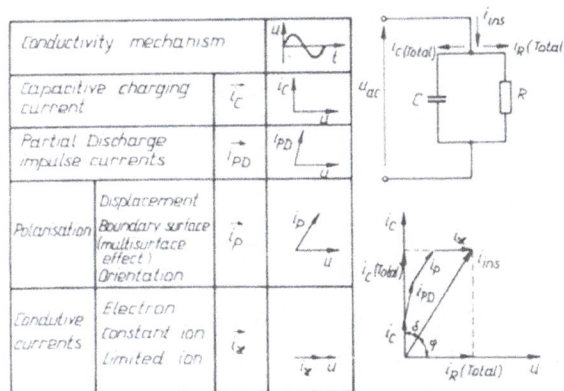

Conductive mechanisms in insulating materials for alternating voltage with equivalent circuit and vector diagrams.

Besides the capacitive charging currents, real or active currents are also present in the dielectrics. These currents are caused due to different types of conductivities and polarizations present in the materials. As shown, these currents not only depend upon the frequency and magnitude of the applied voltage but also upon the thermal conditions of the dielectric. The conductive currents present in an insulating material determine its dielectric power loss property. Each conductive current mechanism causes currents of different characteristics. These currents contribute to the total dielectric current 'i_{ins}' ,as depicted schematically for alternating voltage.

The dielectric loss tangent 'tanδ' is defined as the quotient of active to reactive power loss in a capacitor or in a volume of dielectric. It is derived from the above figure as follows:

$$\tan \delta = \frac{\text{Active Power}}{\text{Reactive Power}} = \frac{u.i_{ins} \cos \varphi}{u.i_{ins} \sin \varphi} = \frac{i_{R(Total)}}{i_{c(Total)}}$$

The dielectric loss tangent (tanδ) represents the complete power loss in a dielectric; hence it is a parameter with which the power losses in a capacitor can be estimated. However, tanδ is a function of frequency and magnitude of the applied voltage as well as the temperature of the dielectric, because these affect the conductivity κ and the polarization processes in the dielectrics. When alternating voltage of rms magnitude U and frequency ω is applied to a condenser having total effective capacitance C, the total capacitive conductive current I_c (Total) is given by

$$I_{c(Total)} = \frac{U}{1/\omega C} = \omega.C.U$$

The active part of the total insulation current is (ωCU .tanδ), hence the active power loss ' P_{ac} ' is given by,

$$P_{ac} = \omega.C.U^2.\tan \delta$$

where ω = 2 π f and if f is in Hz, C in F and U in volts, P is given in Watts. In practice, the capacitance C of the given test object is generally measured. Considering a parallel plate condenser of area A and gap distance d having a dielectric with relative permittivity ε_r in a uniform field E, Equation above can be rewritten as,

$$P_{ac} = \omega.\varepsilon_0 \varepsilon_r \frac{A}{d} (E.d)^2 \tan \delta$$

Or

$$P_{ac} = \varepsilon_0.\omega.E^2.V.\varepsilon_r \tan \delta$$

where V, the volume of the dielectric, is given by (Ad) in this case. Equation shows that the dielectric losses depend upon the applied field intensity and its frequency, volume of the dielectric, its relative permittivity and the loss tangent.

On applying a direct voltage to the dielectric, the losses depend only upon the magnitude of the applied voltage U_{dc} and the dc resistance 'R_{dc}' offered by the dielectric, determined by Equations as follows:

$$P_{dc} = \frac{U^2_{dc}}{R_{dc}}$$

Or

$$P_{dc} = E^2 . V K_{dc}$$

Variation of loss tangent of transformer oil having different moisture contents, measured for a wide range of temperatures by Holle [3.2] is shown in the figure below. These measurements were conducted on a standard test cell (guard ring capacitor) applying a constant 50 Hz voltage.

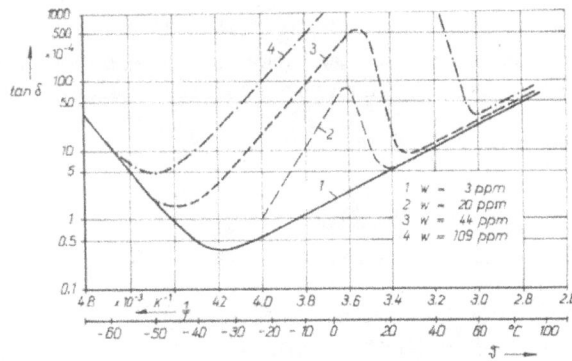

Loss tangent 'tan δ' of transformer oil having different ppm moisture contents
with varying temperature at 50 Hz constant voltage, Holle [3.2].

The loss tangent is a function of the applied voltage. Increase in tanδ of a transformer oil sample, measured at constant temperatures and 50 Hz increasing voltage/field intensity magnitudes is shown in the figure below (Beyer et al). These measurements were made at Schering Institute of HV Engineering, University of Hannover. Extra care was taken to design the measuring arrangement which was developed to be PB free for upto 200 kV. As seen from these curves, the increase in temperature as well as moisture content in oil, both result in a premature increase in tanδ values with increasing voltage.

Figure below shows the variation of tanδ and ε_r of synthetic insulating liquid 'Clophen A 50', with respect to temperature measured at 50Hz constant voltage.

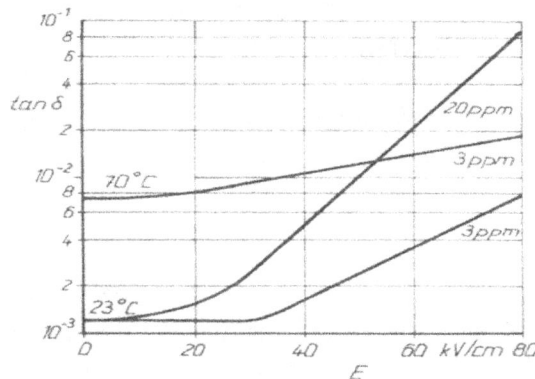

Loss tangent (tan δ) of a transformer oil measured with increasing voltage/field intensity (50 Hz)
at different constant temperatures and for different moisture contents in oil, Bayer et al. [3.6]

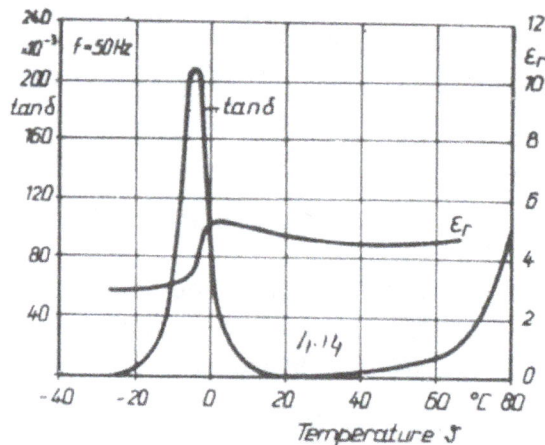

Variation of measured values of tan δ and εr with temperature for synthetic
insulating liquid Clophen A 50 at 50 Hz constant ac Voltage .

Partial Breakdown in Dielectrics

Partial Breakdown (PB) are localized electrical breakdown within an insulation system, restricted only to a certain part of the dielectric. In other words, PB are the discharges which do not bridge the electrodes. PB may occur in gaseous, liquid, solid or in any combination of insulation system at extremely nonuniform field locations above a particular field intensity. The field intensity above which the PB may occur depends upon the physical conditions of the dielectric and its properties.

The term 'Partial Breakdown' includes a group of local breakdown phenomena; corona breakdown in gaseous dielectrics, internal breakdown at voids or cavities within a liquid or solid dielectrics, and surface discharge also known as 'tracking', which appear on the boundary surface of some solid and liquid dielectrics with air or any other gaseous medium.

Corona or PB in transmission lines in atmospheric air cause power loss which is significant in bad weather (rain) conditions. Besides, it also causes Audible Noise (AN), Electromagnetic Interference (EMI) and generation of ozone. PB in Gas Insulated System (GIS) and liquid dielectrics causes all above effects and also de-generates the enclosed gas or the oil. The damage caused can be said to be reversible to an extent. However, in case of internal breakdown taking place within a solid dielectric, the damage is irreversible. Every breakdown event causes deterioration of the material by the energy impact of high energy electrons or accelerated ions, inflicting electrical and chemical transformations. The actual deterioration is, however, dependent upon the material and the surrounding conditions, but with time it may lead to a permanent damage of the dielectric.

Internal Partial Breakdown

Partial Breakdown in solid dielectrics may take place at the so called weak points. These weak points may be voids, cavities, cuts, foreign particles or protrusions in the dielectric, as shown in Fig. i. This figure represents a longitudinal cross sectional view of an extruded solid dielectric power cable, provided with extruded semi conductive sheathing on the high voltage conductor as well as on the outer surface of the dielectric. With the development of modern technology, careful handling and maintaining extra cleanliness, the presence of voids and foreign particles in extruded, moulded and cast dielectrics have been minimised, but not ruled out. Voids and cavities, which are the most common type of weak points in solid dielectrics, may vary in size from a few μm to a few mm.

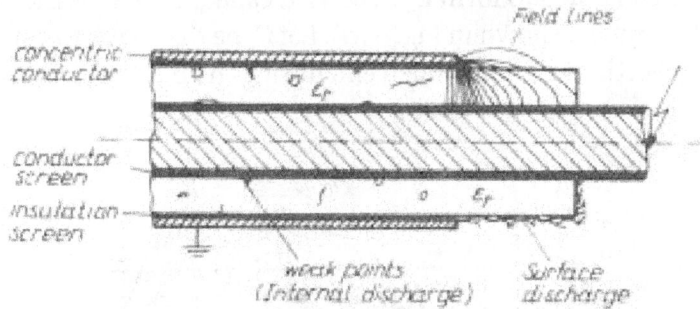

Fig. i A cable cross section showing possibilities of internal and surface discharges

Weak points in general have lower electric strength than the dielectric itself. Still worse is that they acquire higher electric fields. Such weak points can be simulated by a capacitance C_1 within the main dielectric system having capacitance C_3, as shown in Fig. ii (a). If the relative permittivity of a weak point ε_{r1} is lower than that of the dielectric ε_{r2} the electric stress at the weak point may increase. A protrusion in a dielectric acts as sharp extension of the electrode, causing an extreme distortion in the field locally. Thus, the electric stress within the dielectric increases at the weak points considerably.

C_1 - Capacitance of void

C'_2 and C''_2 - Capacitance between void and electrodes

C_3 - Capacitance of the test object

C_c - Coupling capacitor

C_2 - is equivalent of C'_2 and C''_2 in series

Z_m - measuring impendence

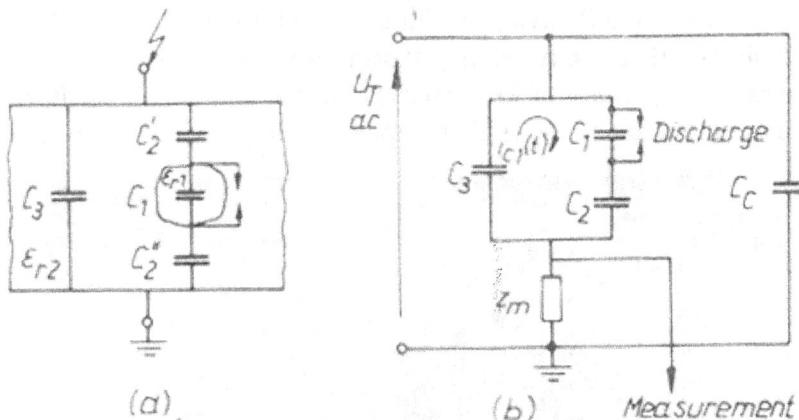

Fig. ii Simulation of internal Partial Breakdown (a) Schematic showing a void forming capacitance
(b) Equivalent circuit diagram for measurement

It is assumed that a capacitance C_1 is formed due to the void within the main dielectric between two electrodes, as shown in Fig. ii (a). Let C_3 be the total capacitance of the electrode system (test object) and ε_{r1} and ε_{r2} the relative permittivities of the void and the main dielectric respectively. The field lines starting and ending at the void form two capacitance C'_2 and C''_2 within the dielectric. C''_2 and C''_2 together in series are combined to form C_2. The void capacitance C_1 is the origin of PB on applying a voltage to the test object U_T, shown in Fig. ii (b). Let C_c be the coupling capacitor required in the circuit to suppress the reflections from the open end during the measurement.

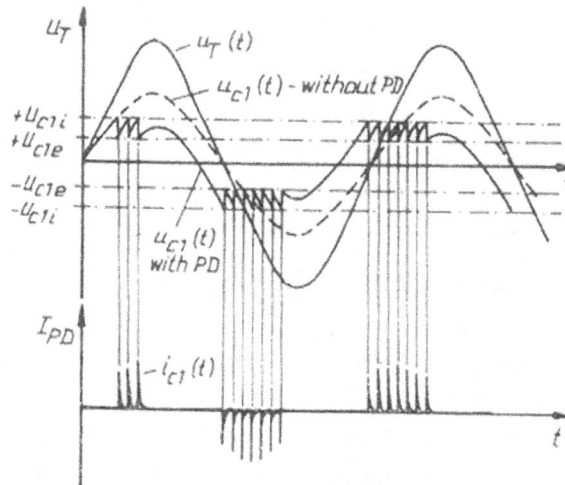

Fig. iii PB voltages and pulse currents at a void or cavity in the dielectric

On applying a 50 Hz, ac test voltage U_T, the capacitance C_1 gets charged. C_1 would experience some proportion of the applied voltage U_T, depending upon the magnitudes of different capacitances. The intensity of field at C_1 will depend upon the shape, size and location, etc. of the void, as explained earlier. On raising the applied voltage further, the inception of breakdown across C_1 appears during the rising part of a half cycle, as shown in Fig. iii, causing the void capacitance C_1 to discharge. The discharge current ic_1 (t), which cannot be directly measured, is a very short pulse current of ns range of duration. The opposite polarity charge, produced by this discharge, is displaced towards the electrodes in the field direction, thus neutralizing the original electric field at the void. If the voltage is still rising at the positive or the negative slope of an ac cycle, new field is built up again, repeating the breakdown phenomenon n number of times during each cycle. The PB current pulses thus produced are measured at the external circuit, giving the intensity of PB in Coulombs. On raising the applied voltage on the test object further, the frequency of occurrence of discharges at C_1 and their intensity increases since the rate of repetition of PB increases. In Fig. iv, the two oscillograms show the PB inception and intensive PB taking place at higher applied voltage on a test object.

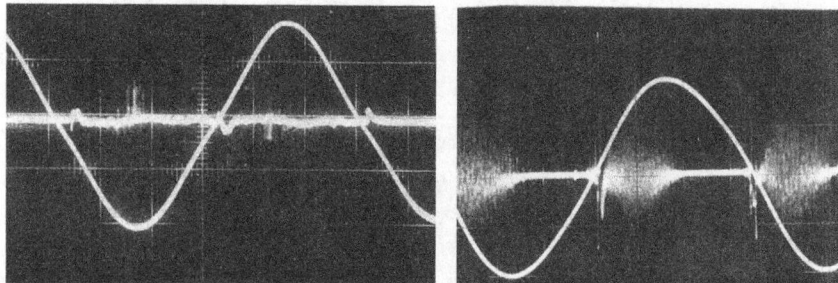

Fig. iv Oscillograms of internal PB in solid dielectric

Assume the magnitude of the voltage at which PD at C1 incept to be U_{C1_i}, and the voltage at which PB extinguish U_{C1_e}, as shown in Fig. iii. The inception voltage U_{C1_i} can be given in terms of the applied voltage and capacitances shown in the equivalent circuit diagram as:

$$U_{C1i} = U_T \frac{C_2}{C_1 + C_2}$$

Due to discharge taking place at C_1, the voltage drop at C_1 may appear to be ΔU c1. Considering no change at C_1 initially, $\Delta Uc1$ is given by,

$$\Delta U_{C1} = U_{C1} \frac{C_2}{C_3 + C_2}$$

If it is assumed that the voltage drop at C_1 equals the corresponding drop in applied voltage at the test object ΔU_T, it can be given by,

$$\Delta U_T = U_T \frac{C_2^2}{(C_3 + C_2)(C_1 + C_2)}$$

The actual charge q_{c1}, transposed to the weak point (void) from the circuit in this process is,

$$q_{c1} = \Delta U_{c1} \left[C_1 + \frac{C_2.C_3}{C_2 + C_3} \right]$$

U_{cii} as well as q_{c1} both are not measurable quantities. However, with the result of potential drop at the test object, a corresponding event takes place at power input terminals. The charge delivered at the power input terminals can be approximated from the equivalent circuit diagram given in Fig. as follows:

$$q \approx \Delta U_T (C_3 + \frac{C_1 C_2}{C_1 + C_2} + C_C)$$

This is a measurable quantity of charge over the measuring impedance Z_m. Since C_1 and C_2 are unknown, the actual inception voltage at the void U_{cii} and the charge magnitude q_{c1} transposed to the void are neither measurable nor calculable. This is the reason that the measurable quantity of charge 'q' at the terminals given by Equation is described by IEC-270 and also IS-6209 as apparent charge 'q_a'. It is defined as follows:

The apparent charge 'q_a' of a partial breakdown is that charge which, if injected instantaneously between the terminals of the test object, would momentarily change the voltage between its terminals by the same amount as the partial breakdown itself. The absolute value I q_a I of the apparent charge is often referred to as the discharge magnitude. The apparent charge is expressed in Coulombs.

The apparent charge q_a defined in this way is not equal to the amount of charge actually transferred across the discharging cavity in the dielectric. It is implied because discharge measuring instruments respond to this quantity. Thus, the magnitude of apparent charge represents the intensity of an individual discharge in a test object. Lemke [3.7] introduced a new concept in this respect in 1975. Since the apparent charge represents the single discharge of maximum intensity taking place in the-test object, Lemke called it 'impulse discharge' denoted by 'q_i' . However in a test object, PB may occur at several locations simultaneously. Therefore, the complete phenomenon occurring in the dielectric is not taken into account by q_i . In order to evaluate all the PB taking place in a test object, Lemke introduced a new concept of 'cumulative discharge 'q_s' (s - sum total). ' q_s' takes into consideration all the individual PB occurring in the test object within a certain time period of the applied voltage. Thus, a more comprehensive investigation of PB phenomenon in an object is made possible by measuring q_s. In Fig. v, variation of measured values of q_i and q_s with increasing voltage are illustrated. These curves measured by Lemke [3.7], are for a needle-plane electrode configuration in air, where the PB taken place at the negative half cycle of the 50 Hz ac voltage have been recorded. Similar curves are also measured on test objects with solid dielectrics as insulation.

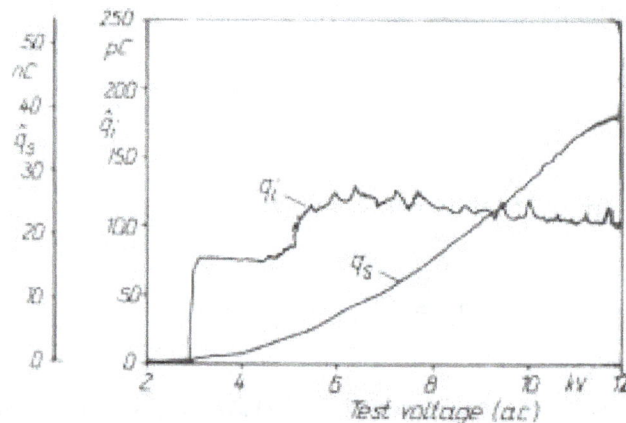

Fig v Variation of impulse 'q_i' and cumulative 'q_s' discharge magnitudes
with increasing voltage, Lemke [3.7].

Besides apparent charge, the other important PB quantities for a test object are the PB inception and extinction voltages. These quantities, as defined in IEC-270, are as follows:

PD Inception Voltage 'U_i'

It is the lowest terminal voltage at which a discharge due to partial breakdown exceeding a specified intensity is observed under specified conditions, when the voltage applied to the test object is gradually increased from a lower value at which no such discharges are observed.

PD Extinction Voltage 'U_e'

It is the voltage at which PD exceeding a specified intensity cease under specified conditions when the voltage is gradually decreased from a value exceeding the inception voltage. The other quantities related to individual discharge are defined as follows:

Repetition Rate 'n'

The PB pulse repetition rate 'n' is the average number of pulses per second.

Energy of an Individual PB, 'w'

The energy due to PB 'w' is the energy dissipated during one individual discharge. It is expressed in joules. Among integrated quantities, the average discharge current I is defined as the sum of the rectified charge quantities passing through the terminals of a test object due to discharges taking place during a certain time interval Δt, given by,

$$\overline{I} = \frac{1}{\Delta t}\left[\left|q_1\right| + \left|q_2\right| + q_3\left|\ldots\ldots\left|q_n\right|\right.\right]$$

The measurable PB quantities described above have been explained with the help of a schematic given by Lemke [3.8], as shown in Fig. vi. Measurement of these quantities on a test object may be able to provide an evaluation of the quality of finish of the dielectric. Satisfactory performance and the life expectancy of an electrical apparatus depend upon the dangerousness of the weak points in the dielectric. Limiting values of PD quantities are standardised for different products in order to ensure quality of finish of the dielectric on an apparatus.

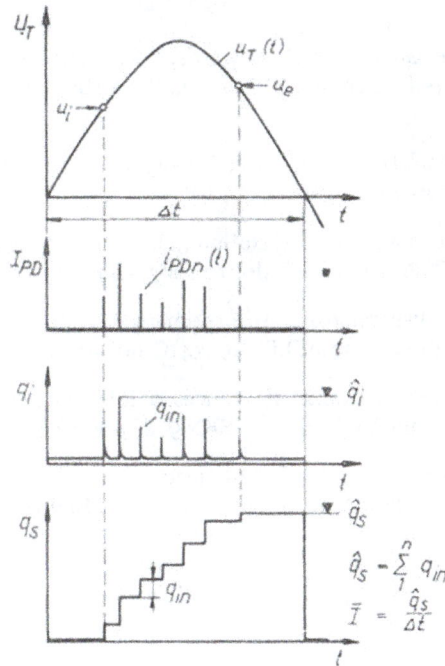

Fig vi

u_i - PB inception , u_e - PB extinction,

Δt - time interval of measurement,

q_i - apparent impulse discharge, q_s - apparent cumulative discharge,

$\hat{q}_j$ - impulse discharge of maximum intensity,

$\hat{q}_s$ - sum total of individual discharge in a time interval of Δt,

$\overline{I}$ - average discharge current.

References

- Ravindra Arora; Wolfgang Mosch (25 February 2011). High Voltage and Electrical Insulation Engineering. John Wiley & Sons. pp. 249–. ISBN 978-1-118-00896-6. Retrieved 9 January 2012

- Cole, Kenneth S, Robert H (1941). "Dispersion and Absorption in Dielectrics: I - Alternating Current Characteristics". Journal of Chemical Physics. 9: 341–351

- Havriliak, S.; Negami, S. (1967). "A complex plane representation of dielectric and mechanical relaxation processes in some polymers". Polymer. 8: 161–210. doi:10.1016/0032-3861(67)90021-3

- Gaseous dielectrics with low global warming potentials – US Patent Application 20080135817 Description Archived October 13, 2012, at the Wayback Machine.. Patentstorm.us (2006-12-12). Retrieved on 2011-08-21

- Safari, Ahmad (2008). Piezoelectric and acoustic materials for transducer applications. Springer Science & Business Media. p. 21. ISBN 0387765409

- Cole, K.S.; Cole, R.H. (1942). "Dispersion and Absorption in Dielectrics - II Direct Current Characteristics". Journal of Chemical Physics. 10: 98–105. Bibcode:1942JChPh..10...98C. doi:10.1063/1.1723677

- Hilfer, J. (2002). "H-function representations for stretched exponential relaxation and non-Debye susceptibilities in glassy systems". Physical Review E. 65: 061510. Bibcode:2002PhRvE..65f1510H. doi:10.1103/physreve.65.061510

- Aggarwal, M.D.; A.K. Batra; P. Guggilla; M.E. Edwards; B.G. Penn; J.R. Currie Jr. (March 2010). "Pyroelectric Materials for Uncooled Infrared Detectors: Processing, Properties, and Applications" (PDF). NASA. p. 3. Retrieved 26 July 2013

- Hartmut Haug; Stephan W. Koch (1994). Quantum Theory of the Optical and Electronic Properties of Semiconductors. World Scientific. p. 196. ISBN 981-02-1864-8

- Zorn, R. (2002). "Logarithmic moments of relaxation time distributions". Journal of Chemical Physics. 116 (8): 3204–3209. Bibcode:2002JChPh.116.3204Z. doi:10.1063/1.1446035.

- Cole, K.S.; Cole, R.H. (1941). "Dispersion and Absorption in Dielectrics - I Alternating Current Characteristics". J. Chem. Phys. 9: 341–352. Bibcode:1941JChPh...9..341C. doi:10.1063/1.1750906

- Gorenflo, Rudolf; Kilbas, Anatoly A.; Mainardi, Francesco; Rogosin, Sergei V. (2014). Springer, ed. Mittag-Leffler Functions, Related Topics and Applications. ISBN 978-3-662-43929-6

- Lyon, David (2013). "Gap size dependence of the dielectric strength in nano vacuum gaps". IEEE Transactions on Dielectrics and Electrical Insulation. 20 (4). doi:10.1109/TDEI.2013.6571470

Liquid and Solid Dielectrics

Solid dielectric materials such as porcelain, mica, plastics, etc. are used when current is transferred within electrical circuits containing different voltages. Liquid dielectrics on the other hand are used to prevent electrical discharge. The topics discussed in the chapter are of great importance to broaden the existing knowledge on liquid and solid dielectrics.

Liquid Dielectric

A liquid dielectric is a dielectric material in liquid state. Its main purpose is to prevent or rapidly quench electric discharges. Dielectric liquids are used as electrical insulators in high voltage applications, e.g. transformers, capacitors, high voltage cables, and switchgear (namely high voltage switchgear). Its function is to provide electrical insulation, suppress corona and arcing, and to serve as a coolant.

A good liquid dielectric should have high dielectric strength, high thermal stability and chemical inertness against the construction materials used, non-flammability and low toxicity, good heat transfer properties, and low cost.

Liquid dielectrics are self-healing; when an electric breakdown occurs, the discharge channel does not leave a permanent conductive trace in the fluid.

The electrical properties tend to be strongly influenced by dissolved gases (e.g. oxygen or carbon dioxide, dust, fibers, and especially ionic impurities and moisture. Electrical discharge may cause production of impurities degrading the dielectric's performance.

Some examples of dielectric liquids are transformer oil, perfluoroalkanes, and purified water.

Common Liquid Dielectrics

Name	Dielectric constant	Max. breakdown strength (MV/cm)	Properties
Mineral oil		1.0	Flammable. Common type of transformer oil.
n-Hexane		1.1-1.3	Flammable. Used in some capacitors.
n-Heptane			Flammable.
Castor oil	4.7		High dielectric constant. Flammable. Refined and dried castor oil is used in some high voltage capacitors.
Silicone oil		1.0-1.2	More expensive than hydrocarbons. Less flammable.
Polychlorinated biphenyls			Formerly used in transformers and capacitors. Persistent organic pollutant, toxic, now phased out. Low flammability.

Name	Dielectric constant	Max. breakdown strength (MV/cm)	Properties
Purified water	~80		High thermal capacity, good cooling properties. Low electrical conductivity when free of ions.
Benzene		1.1	Toxic, flammable.
Liquid oxygen		2.4	Cryogenic. Highly flammable with combustible materials.
Liquid nitrogen	1.43	1.6-1.9	Cryogenic. Used as coolant with many low-temperature sensors and high-temperature superconductors.
Liquid hydrogen		1.0	Cryogenic. Flammable.
Liquid helium		0.7	Cryogenic. Used with superconductors.
Liquid argon		1.10-1.42	Cryogenic.

From the point of view of molecular arrangements, liquids can be described as 'highly compressed gases ' in which the molecules are very closely arranged. This is known as kinetic model of the liquid structure. However, the movement of charged particles, their microscopic streams and interface conditions with other materials cause a distortion in the otherwise undisturbed molecular structure of the liquids. The well known terminology describing the breakdown mechanisms in gaseous dielectrics, such as, impact ionization, mean free path, electron drift etc. is, therefore, also applicable for liquid dielectrics.

Liquid dielectrics are accordingly classified in between the two states of matter, that is gaseous and solid insulating materials. This intermediate position of liquid dielectrics is also characterized by its wide range of application in power apparatus. Insulating oils are used in power and instrument transformers, power cables, circuit breakers, power capacitors etc. Here they perform a number of functions simultaneously, namely:

- insulation between the parts carrying voltage and the grounded container, as in transformers.

- impregnation of insulation provided in thin layers of paper or other materials, as in Transformers, cables and capacitors, where oils or impregnating compounds are used.

- cooling action by convection in transformers and oil filled cables through circulation.

- filling up of the voids to form an electrically stronger integral part of a composite dielectric.

- arc extinction in oil circuit breakers.

- high capacitance provided by liquid dielectrics with high permittivity to power capacitors.

A large number of natural and synthetic liquids are available which can be used as dielectrics. These possess a very high electric strength and their viscosity and permittivity vary in a wide range. The appropriate application of a liquid dielectric in an apparatus is determined by its physical, chemical and electrical properties on one hand, and on the other, it depends upon the requirements of the functions to be performed.

Classification of Liquid Dielectrics

Dielectric materials can be divided into two broad classifications: organic and inorganic. Organic dielectrics are basically chemical compounds containing carbon. Earlier, under organic chemistry only those compounds which were derived from either plant or animal organism were considered. This concept underwent changes with concurrent developments in chemical technology. Carbon compounds, in general, are now called organics. Among the main natural insulating materials of this type are: petroleum products and mineral oils. The most important and widely used organic liquid dielectrics for electrical power equipment are mineral oils. The other natural organic insulating materials are asphalt, vegetable oils, wax, natural resins, wood and fibre plants (fibrins).

A large number of synthetic organic insulating materials are also produced. These are nothing but substituents of hydrocarbons in gaseous or liquid forms. In gaseous forms are fluorinated and chlorinated carbon compounds. Their liquid forms are chlorinated diphenyles, besides some nonchlorinated synthetic hydrocarbons. The chlorodiphenyles, although possessing some special properties, are not widely used because of being unfriendly to human beings and very costly.

Among halogenfree synthetic oils are polymerization products, the polyisobutylenes and the siliconpolycondensates. Polyisobutylene offers, better dielectric and thermal properties than mineral oils for its application in power cables and capacitors, but it is many times more expensive. Silicon oils are top grade, halogenfree synthetic insulating liquids. They have excellent stable properties, but because of being costly, have so far found limited application in power apparatus for special purpose.

Among inorganic liquid insulating materials, highly purified water, liquid nitrogen, oxygen, argon, sulphurhexafluoride, and helium etc. have been investigated for possible use as dielectrics. Liquified gases, having high electric strength, are more frequently used in cryogenic applications, whereas, because of its high relative permittivity, low cost, easy handling and disposal, Water and water mixtures are being actively investigated for use as dielectrics (for example, in pulse power capacitors and pulsed power modulators etc.).

To summarize the classification of liquid dielectrics explained above, a schematic is drawn in the following diagram. The insulating liquids, commonly applied in high voltage apparatus, are classified as illustrated in the following diagram.

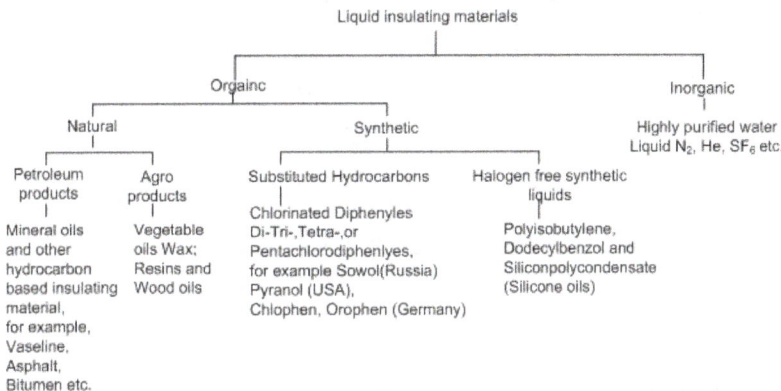

Electric Conduction in Insulating Liquids

The conduction mechanism in a dielectric liquid is strongly affected by the degree of its purity. In liquids, which have not been highly purified, when subjected to fields upto about a few kV/cm the conduction is primarily ionic. Ionic conduction is affected by the dissociation process of impurities as well as the injection of charge by the electrodes through electrochemical reactions. The electrons do not take part in the passage of current through such liquids. A typical characteristic of conduction current density J versus the applied dc voltage U in a pure liquid is shown in Fig. i. At low voltages, the current varies proportionally to the voltage, region 1, representing the ohmic behavior. On raising the applied voltage, hence the field intensity, the ionic current gets saturated, region 2. At still higher applied voltage the current density increases rapidly until the breakdown occurs, region 3.

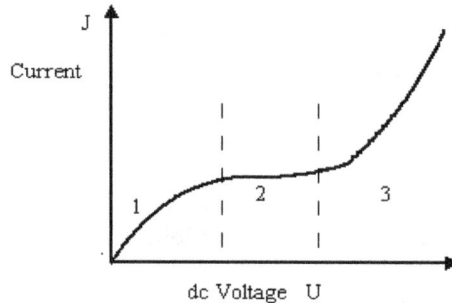

Fig. i Typical voltage-current (U-J) characteristic in dielectric liquids.

Similar behavior is observed generally in, both, polar and nonpolar liquids. In some cases the saturation region may be completely missing. This characteristic has an analogy to what has also been observed commonly in gases.

In Fig. ii the conduction current characteristics are shown for negative direct voltage applied to a needle-plane electrode configuration in transformer oil and liquid nitrogen, measured by Takashima et al. [3.1] for different radii of curvature 'r_t' of the needle tip. For these experiments, the needle constituted of a polished platinum wire of radius 0.25 mm and the plane of 50 mm diameter Rogowsky profile curved plate of stainless steel. The current was measured through the plane electrode for a constant gap distance of 4 mm. The sharp bend measured on the curve is supposed to represent the development of a strong space charge around the needle.

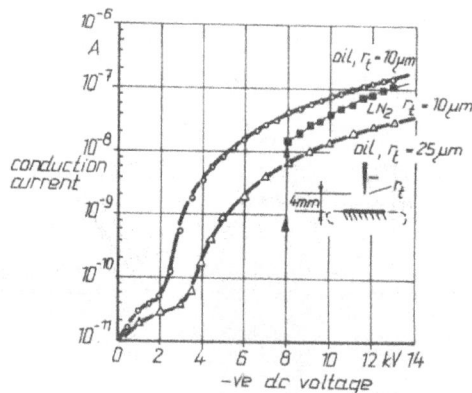

Fig. ii Conduction current in oil and liquid nitrogen for negative needle-plane electrode gap of 4mm with increasing voltage. Oil at 20°C and LN2 at 77.3 K Takashima et al. [3.1].

The concentration of charge carriers and the viscosity of insulating liquids are both strongly affected by the temperature which in turn influences the conductivity of the liquid. According to Van't Hoffsch law, the conductivity 'κ' within a certain range of temperature follows the relation:

$$K = K_0 \exp(-f/kT)$$

where k is Boltzmann constant, T absolute temperature, κ_0 and F are material constants. F is known as 'activation energy' and is expressed as kcal/mole.

However, Van't Hoffsch law is valid only in the region where the conduction current follows the ohmic behavior, that is region 1 in Fig. i. It thus depends upon the particular liquid and its impurity contents (for example, humidity content etc.).Transformer oil conductivity variation for a wide range of temperature for different moisture contents measured by Holle [3.2] are illustrated in Fig. iii. It is evident from this figure that the conductivity of oil increases as the water ppm content rises.

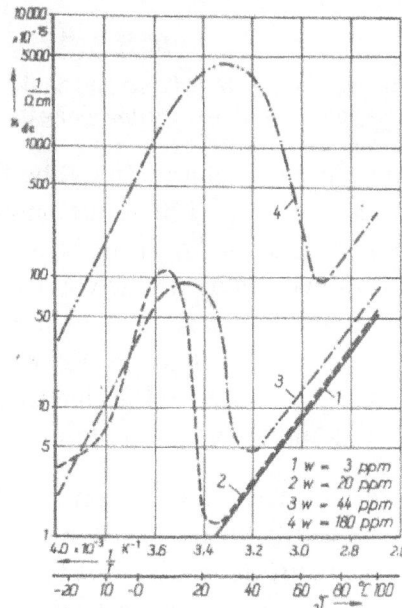

Fig. iii Direct current conductivity of transformer oil for different water contents w (ppm) with respect to the reciprocal of absolute temperature, Holle [3.2].

In case of highly purified liquids which are able to withstand high field intensity, the conduction phenomenon is different. By increasing the field intensity to the order of a few hundred kV/cm, the conduction current in both polar and nonpolar liquids is augmented predominantly by charge carriers injected into the liquid from the electrode surfaces. Besides, the field aided dissociation process of molecules is intensified. Under certain given conditions, the presence of an injected space charge, say of density q, gives rise to a Coulomb's force of density qE. Due to the action of this force, hydrodynamic instability is caused, developing convection motion in the liquid. Investigations made by many authors, on both polar and nonpolar liquids, have concluded that whenever conduction in an insulating fluid is accompanied with a significant charge injection, the convection motions occur, which are also known as 'electrohydrodynamic' (EHD) motions. The EHD motions augment the passage of current depending upon the charge injection intensity, which in turn is determined by the applied voltage, the nature of liquid and the electrode material, Zahn [3.3], Takashima [3.1] and Atten [3.4].

Atten [3.4] explained the effect of the phenomenon of EHD motion on the conduction in liquid dielectrics under two extreme conditions: with and without unipolar injection of charge. For his experimentation, an extremely nonuniform field configuration between a knife and a plane was taken, as shown in Fig. iv.

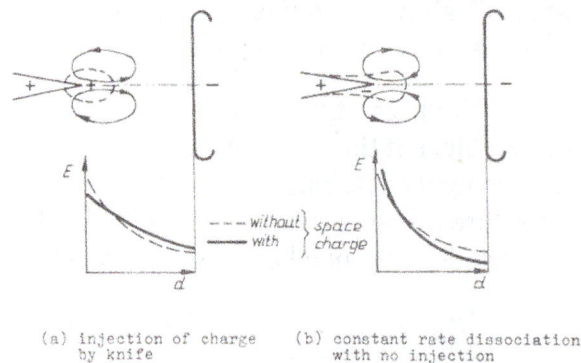

(a) injection of charge   (b) constant rate dissociation
    by knife   with no injection

Fig. iv Schematic representation of space charge zone and liquid motion streams between knife-plane electrodes and the field distribution along the axis, Atten [3.4].

Consider first the case of injection of unipolar charge (ions) by the knife, Fig. iv (a). The field at the tip of the knife electrode is reduced because of the same polarity space charge being injected, whereas an increase in the field intensity is caused over the basic field towards the plane electrode. The space charge zone is of finite extent when a moderate voltage is applied. Under this condition, the liquid relaxation time is lower than the transit time of the ions from the knife to the plane. The total current is, therefore, higher than the ohmic current because of the enhancement of field intensity near the plane. Moreover, due to the action of the electric field on the same polarity charge close to the knife, a 'stream' or 'jet' of liquid towards the plane is induced in this region.

In case of pure conduction with no injection of charge, Fig. iv (b), an opposite polarity space charge near the knife extends further. Consequently, the space charge field at the tip increases, and is reduced with respect to the basic field towards the plane. This results in lower total current than the ohmic current (sub-ohmic behavior). The opposite polarity space charge in the vicinity of the knife gives rise to a liquid motion directed towards the knife.

At extremely high field intensities (1000 kV/cm and above) approaching the 'intrinsic strength' of liquid dielectrics, the field emission at the electrode surfaces affects the conduction phenomenon. Under these conditions, the conduction current is dominated by electrons.

Classification of Solid Dielectrics

A vast number of solid insulating materials are used in electrical power engineering. With the invention of modem insulating materials, many earlier conventional materials have been discarded, especially in case of electrical machines, capacitors and power cables. From the usage point of view, the solid insulating materials can be divided into the following three broad categories:

Moulding materials: These are used for providing mechanically rigid forms of insulation, for example, insulators and bushings etc. Such materials are required to have good electrical insulating properties besides having high mechanical strength. These are usually made out of ceramics, glass (toughened glass), fibre glass reinforced plastics and epoxyresins.

Jacketing materials: Polymers have been found suitable for providing insulating jackets to the conductors. For example, polyethylene (PE), polyvinylchloride (PVC), natural and synthetic rubber and paper used in power cables, capacitors and transformers. Mica and fibre glass based plastic tapes are used in electrical machines.

Filling materials: Beside oils, wax based draining land non draining impregnating compounds of different types are used to impregnate paper in power cables, transformers, capacitors, and instrument transformers.

Insulating mechanical support: In the form of plates, pipes and ledges insulating supports are required in transformers, GIS, circuit breakers and isolators. The products, such as press boards, hard paper (thin paper laminates), wood (yellow teak) are used in transformers. Ebonite, also known as vulcanite, is a form of cross linked rubber containing large proportion of sulphur, bakelite, a hard synthetic material, and acrylic resin plates are used for insulating supports in other equipment.

However, the most common method of classifying the solid insulating materials is based upon their chemical compositions, distinguished mainly between inorganic and organic materials.

Among the inorganic materials; ceramics, glass, fibre glass, enamel, mica and asbestos are important and have found their wide application as dielectrics in the order of the mentioned sequence. They distinguish themselves in their unique ability to withstand high temperatures in addition to their being highly chemical resistant. There is practically negligible sign of ageing in these materials. However, it is difficult to machine or process them. These materials are basically inhomogeneous both microscopically as well as macroscopically.

Solid organic materials used in electrical engineering are: paper, wood, wax, leather, cotton besides a number of natural and synthetic resins, rubbers and plastics, also known as polymers. Wood being so hygroscopic, is rarely used as such, but wooden poles supporting overhead cables and distribution lines are being widely used in North America.

The polymeric organic insulating materials used in electrical engineering have a very high molecular weight and consist of two or more polymeric compounds of several structural units normally bound together by covalent bonds. The individual structural units may consist of single atoms or may be molecular in nature, which repeat in a regular order.

A polymer can be obtained by the reaction of compounds having at least two reactive functional groups which can react under suitable conditions. These chemical compounds are known as 'monomers'. There can be one or a mixture of reacting monomers and in the latter case, the resulting material is called 'copolymer'.

There are several ways of classifying polymers and one of them is by their response to heat. Accordingly, the polymers are divided into two groups of materials as follows.

Thermoplastic Polymers

Thermoplastic polymers soften and supple on heating and 'solidify' on cooling. The heating and cooling cycle within a certain temperature limit can be applied to these materials several times without affecting their properties. Polymers which generally have a linear structure fall in this category. The synthetic thermoplastic materials. used in electrical engineering for insulation purpose are; polyethylene (PE), polyvinylchloride (PVC), polypropylene (PP) and polyamide (PA). The extent of their application in the industry is in the order of their above mentioned sequences. Such materials have relatively poor thermal resistance and their properties deteriorate rapidly at higher temperature.

Thermoset Polymers

The polymers which soften when heated for the first time resulting into cross-linking reaction (network formation), are known as thermoset polymers. This reaction, leading to the formation of network structure, is also known as curing or setting of the polymer. All monomers having functionality equal to two would give rise to linear polymers and, therefore, thermoplastic materials. However, when they have functionality more than two, the resulting polymer has a network structure. The polymers used for electrical insulation purpose are desired to retain their rubbery (flexible) properties. Hence, these are defined as 'lightly cross-linked polymers'.

The chemical reaction leading to cross-linking is achieved with the help of an additive, known as 'agent'. The term 'cross-linking agent' is very general. It refers to (a) molecules which bridge two polymer molecules, for example, vulcanizing agents; such as sulphur, selenium and sulphur monochloride for rubbers; and polyamines in expoxide resins, (b) molecules which initiate a cross-linking reaction, for example, peroxides which initiate a double bond polymerisation in polyesters and dicumyl peroxide in low density polyethylene (LDPE); (c) those which are purely catalytic in their action such as, acids for phenolic resins, amines in epoxides; and (d) active site generators, for example, peroxides which abstract protons from the polymer chains.

The thermal, mechanical and electrical properties of different cross-linked materials vary considerably from each other. However, in general, after crosslinking a material does not soften on reheating, instead it becomes thermally more stable compared to thermoplastic materials. Cross-linked polymer resins are, for example; polyester-resin, phenolresin, siliconresin and the most widely used in electrical engineering are the expoxyresins. Among rubbers, typical examples used for electrical insulation are; natural rubber also known as 'India rubber', and a large number of synthetic rubbers like silicon rubber (SiR), ethylene-propylene rubber (EPR), etc.

The bulk properties of a polymer can be altered suitably by incorporation of a number of additives. Variations in the choice of additives can produce widely differing products. This is true in particular with PVC. In terms of functions, there are a large number of groups of additives, of which the following most important are commonly used in preparing polymer compounds for electrical insulation purpose:

1. Fillers-usually applied to modify physical properties, mainly mechanical, of a polymer.

2. Plasticisers and Softeners-to lower the melt viscosity and also to change physical properties (softness, flexibility).

3. Colorants-normally soluble colorants(dyestuffs) are used.

4. Anti-ageing additives to prevent structural degradation due to chemical reactions like oxidation, ozone attack, dehydroclorination (especially with PVC) and ultra-violet attack, e.g. sunlight.

5. Flame retarders - to improve the degree of fire resistance of polymers.

6. Cross-linking agents-to achieve intermolecular combination at the chain ends.

Composite Insulating Systems

Some very high quality solid insulating systems have been produced by combining different dielectric materials. Common composites of organic and inorganic materials are; fibre glass reinforced plastics, mica based plastic tapes, quartz, fibre and mica mixed with synthetic resins, such as epoxyresin used in electrical machines, etc. Oil and wax based compound impregnated insulation systems with paper and polymer tapes are commonly applied in power cables, capacitors and transformers. 'Uniaxially oriented polyethylene' (UOPE) tape, developed recently has been found to have very good compatibility with oils besides having good mechanical properties. In such composite insulating systems, the advantages of inherent properties of their constituent materials are made use of. Thus, it is possible to create insulating systems having better thermal, mechanical and electrical properties.

Nano-composite Dielectrics

Research worldwide has been able to document significant improvement in the electrical and other properties of polymer composites through incorporation of nanoparticulates. Research in polymer processing is necessary to study the functionalization of the particulate surfaces to provide preferential coupling to the host polymer in the utilization of the emerging breed of new dielectric material. The nano-composites are intrinsically different in that they appear to be dominated by the characterstics of the internal interfaces. These properties appear to arise when the infilled material has a similar length scale to that of the polymer chain. Attempts to engineer nanodielectrics by changing the conditions at the interface do suggest that indeed some degree of tailoring of dielectric properties and self assembly may be possible. In table below, examples of nano-composite insulating materials currently under investigation are given.

Table: Examples of nano-composite systems under investigations.

Base polymers	Nano-materials
Polyolephins	Clays
Epoxies/phenolics	Inorganic oxides
Elastomers	Carbon nanotubes
Ethylene-vinyl copolymers	Graphite

Permittivity

In electromagnetism, permittivity or absolute permittivity, usually denoted by the Greek letter ε (epsilon), is the measure of resistance that is encountered when forming an electric field in a particular medium. More specifically, permittivity describes the amount of charge needed to generate one unit of electric flux in a particular medium. Accordingly, a charge will yield more electric flux in a medium with low permittivity than in a medium with high permittivity. Thus, permittivity is the measure of a material's ability to resist an electric field, not its ability to 'permit' it (as the name 'permittivity' might seem to suggest).

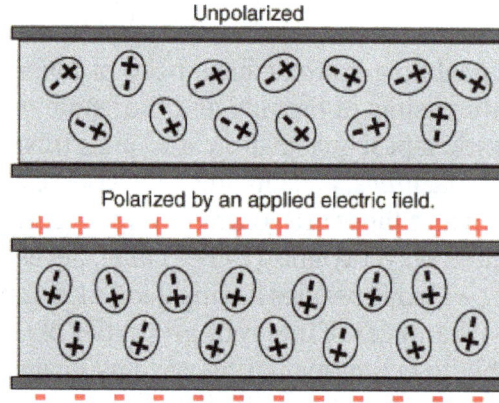

A dielectric medium showing orientation of charged particles creating polarization effects. Such a medium can have a lower ratio of electric flux to charge (more permittivity) than empty space.

The SI unit for permittivity is Farad per meter (F/m or F·m⁻¹).

The lowest possible permittivity is that of a vacuum. Vacuum permittivity, sometimes called the electric constant, is represented by ε_0 and has a value of approximately 8.85×10^{-12} F/m.

The permittivity of a dielectric medium is often represented by the ratio of its absolute permittivity to the electric constant. This dimensionless quantity is called the medium's relative permittivity (ε_r) or dielectric constant (κ).

$$\kappa = \varepsilon_r = \frac{\varepsilon}{\varepsilon_0}$$

By definition, a perfect vacuum has a relative permittivity of exactly 1. The difference in permittivity between a vacuum and air can often be considered negligible, as $\kappa_{air} = 1.0006$.

Relative permittivity is directly related to electric susceptibility (χ), which is a measure of how easily a dielectric polarizes in response to an electric field, given by

$$\chi = \kappa - 1$$

otherwise written as

$$\varepsilon = \varepsilon_r \varepsilon_0 = (1+\chi)\varepsilon_0$$

Units

The standard SI unit for permittivity is Farad per meter (F/m or F·m⁻¹).

$$\frac{F}{m} = \frac{C}{V \cdot m} = \frac{C^2}{N \cdot m^2} = \frac{A^2 \cdot s^4}{kg \cdot m^3} = \frac{N}{V^2}$$

Explanation

In electromagnetism, the electric displacement field D represents how an electric field E influences the organization of electric charges in a given medium, including charge migration and electric dipole reorientation. Its relation to permittivity in the very simple case of *linear, homogeneous, isotropic* materials with *"instantaneous" response* to changes in electric field is

$$\mathbf{D} = \varepsilon\mathbf{E}$$

where the permittivity ε is a scalar. If the medium is anisotropic, the permittivity is a second rank tensor.

In general, permittivity is not a constant, as it can vary with the position in the medium, the frequency of the field applied, humidity, temperature, and other parameters. In a nonlinear medium, the permittivity can depend on the strength of the electric field. Permittivity as a function of frequency can take on real or complex values.

In SI units, permittivity is measured in farads per meter (F/m or $A^2 \cdot s^4 \cdot kg^{-1} \cdot m^{-3}$). The displacement field D is measured in units of coulombs per square meter (C/m^2), while the electric field E is measured in volts per meter (V/m). D and E describe the interaction between charged objects. D is related to the *charge densities* associated with this interaction, while E is related to the *forces* and *potential differences*.

Vacuum Permittivity

The vacuum permittivity ε_0 (also called permittivity of free space or the electric constant) is the ratio $\dfrac{\mathbf{D}}{\mathbf{E}}$ in free space. It also appears in the Coulomb force constant,

$$k_e = \frac{1}{4\pi\varepsilon_0}$$

Its value is

$$\varepsilon_0 \overset{\text{def}}{=} \frac{1}{c_0^2\mu_0} = \frac{1}{35950207149.4727056\pi} \text{ F/m} \approx 8.8541878176...\times10^{-12} \text{ F/m}$$

where

- c_0 is the speed of light in free space,

- μ_0 is the vacuum permeability.

The constants c_0 and μ_0 are defined in SI units to have exact numerical values, shifting responsibility of experiment to the determination of the meter and the ampere. (The approximation in the second value of ε_0 above stems from π being an irrational number).

Relative Permittivity

The linear permittivity of a homogeneous material is usually given relative to that of free space, as a relative permittivity ε_r (also called dielectric constant, although this sometimes only refers to the static, zero-frequency relative permittivity). In an anisotropic material, the relative permittivity may be a tensor, causing birefringence. The actual permittivity is then calculated by multiplying the relative permittivity by ε_o:

$$\varepsilon = \varepsilon_r \varepsilon_0 = (1+\chi)\varepsilon_0,$$

where χ (frequently written χ_e) is the electric susceptibility of the material.

The susceptibility is defined as the constant of proportionality (which may be a tensor) relating an electric field E to the induced dielectric polarization density P such that

$$\mathbf{P} = \varepsilon_0 \chi \mathbf{E},$$

where ε_o is the electric permittivity of free space.

The susceptibility of a medium is related to its relative permittivity ε_r by

$$\chi = \varepsilon_r - 1.$$

So in the case of a vacuum,

$$\chi = 0.$$

The susceptibility is also related to the polarizability of individual particles in the medium by the Clausius-Mossotti relation.

The electric displacement D is related to the polarization density P by

$$\mathbf{D} = \varepsilon_0 \mathbf{E} + \mathbf{P} = \varepsilon_0 (1+\chi)\mathbf{E} = \varepsilon_r \varepsilon_0 \mathbf{E}.$$

The permittivity ε and permeability μ of a medium together determine the phase velocity $v = \dfrac{c}{n}$ of electromagnetic radiation through that medium:

$$\varepsilon\mu = \frac{1}{v^2}.$$

Practical Applications

Determining Capacitance

The capacitance of a capacitor is based on its design and architecture, meaning it will not change with charging and discharging. The formula for capacitance is given as

$$C = \varepsilon \frac{A}{d}$$

where A is the area of one plate, d is the distance between the plates, and ε is the permittivity of the medium between the two plates. For a capacitor with dialectric constant κ, it can be said that

$$C = \kappa \, \varepsilon_0 \frac{A}{d}$$

Gauss' Law

Permittivity is connected to electric flux (and by extension electric field) through Gauss' Law. Gauss' Law states that for a closed Gaussian surface,

$$\Phi_E = \frac{Q_{enc}}{\varepsilon_0} = \oint_s E \cdot dA$$

where Φ_E is the net electric flux passing through the surface, Q_{enc} is the charge enclosed in the Gaussian surface, E is the electric field vector at a given point on the surface, and dA is a differential area vector on the Gaussian surface.

If the Gaussian surface uniformly encloses an insulated, symmetrical charge arrangement, the formula can be simplified to

$$EA\cos\theta = \frac{Q_{enc}}{\varepsilon_0}$$

where θ represents the angle between the electric field vector and the area vector.

If all of the electric field lines cross the surface at 90°, the formula can be further simplified to

$$E = \frac{Q_{enc}}{\varepsilon_0 A}$$

Because the surface area of a sphere is $4\pi r^2$, the electric field a distance r away from a uniform, spherical charge arrangement is

$$E = \frac{Q}{\varepsilon_0 A} = \frac{Q}{\varepsilon_0 (4\pi r^2)}$$

$$E = \frac{Q}{4\pi\varepsilon_0 r^2} = \frac{kQ}{r^2}$$

where k is Coulomb's constant ($\sim 9.0 \times 10^9 \ m/F$). This formula applies to the electric field due to a point charge, outside of a conducting sphere or shell, outside of a uniformly charged insulating sphere, or between the plates of a spherical capacitor.

Dispersion and Causality

In general, a material cannot polarize instantaneously in response to an applied field, and so the more general formulation as a function of time is

$$P(t) = \varepsilon_0 \int_{-\infty}^{t} \chi(t - t') E(t') dt'.$$

That is, the polarization is a convolution of the electric field at previous times with time-dependent susceptibility given by $\chi(\Delta t)$. The upper limit of this integral can be extended to infinity as well if one defines $\chi(\Delta t) = 0$ for $\Delta t < 0$. An instantaneous response would correspond to a Dirac delta function susceptibility $\chi(\Delta t) = \chi\delta(\Delta t)$.

It is convenient to take the Fourier transform with respect to time and write this relationship as a function of frequency. Because of the convolution theorem, the integral becomes a simple product,

$$\mathbf{P}(\omega) = \varepsilon\ \chi(\omega)\mathbf{E}(\omega).$$

This frequency dependence of the susceptibility leads to frequency dependence of the permittivity. The shape of the susceptibility with respect to frequency characterizes the dispersion properties of the material.

Moreover, the fact that the polarization can only depend on the electric field at previous times (i.e. effectively $\chi(\Delta t) = 0$ for $\Delta t < 0$), a consequence of causality, imposes Kramers–Kronig constraints on the susceptibility $\chi(0)$.

Complex Permittivity

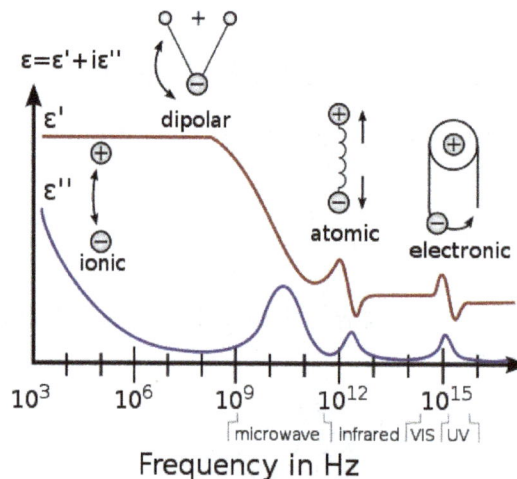

A dielectric permittivity spectrum over a wide range of frequencies. ε' and ε'' denote the real and the imaginary part of the permittivity, respectively. Various processes are labeled on the image: ionic and dipolar relaxation, and atomic and electronic resonances at higher energies.

As opposed to the response of a vacuum, the response of normal materials to external fields generally depends on the frequency of the field. This frequency dependence reflects the fact that a material's polarization does not respond instantaneously to an applied field. The response must always be *causal* (arising after the applied field) which can be represented by a phase difference. For this reason, permittivity is often treated as a complex function of the (angular) frequency ω of the applied field:

$$\varepsilon \rightarrow \hat{\varepsilon}(\omega)$$

(since complex numbers allow specification of magnitude and phase). The definition of permittivity therefore becomes,

$$D_0 e^{-i\omega t} = \hat{\varepsilon}(\omega) E_0 e^{-i\omega t},$$

where

- D_0 and E_0 are the amplitudes of the displacement and electric fields, respectively,

- i is the imaginary unit, $i^2 = -1$.

The response of a medium to static electric fields is described by the low-frequency limit of permittivity, also called the static permittivity ε_s (also ε_{DC}):

$$\varepsilon_s = \lim_{\omega \to 0} \hat{\varepsilon}(\omega).$$

At the high-frequency limit, the complex permittivity is commonly referred to as ε_∞. At the plasma frequency and below, dielectrics behave as ideal metals, with electron gas behavior. The static permittivity is a good approximation for alternating fields of low frequencies, and as the frequency increases a measurable phase difference δ emerges between D and E. The frequency at which the phase shift becomes noticeable depends on temperature and the details of the medium. For moderate field strength (E_0), D and E remain proportional, and

$$\hat{\varepsilon} = \frac{D_0}{E_0} = |\varepsilon| e^{i\delta}.$$

Since the response of materials to alternating fields is characterized by a complex permittivity, it is natural to separate its real and imaginary parts, which is done by convention in the following way:

$$\hat{\varepsilon}(\omega) = \varepsilon'(\omega) + i\varepsilon''(\omega) = \left|\frac{D_0}{E_0}\right| (\cos\delta + i\sin\delta).$$

where

- ε' is the real part of the permittivity, which is related to the stored energy within the medium;

- ε'' is the imaginary part of the permittivity, which is related to the dissipation (or loss) of energy within the medium;

- δ is the loss angle.

The choice of sign for time-dependence, $e^{-i\omega t}$, dictates the sign convention for the imaginary part of permittivity. The signs used here correspond to those commonly used in physics, whereas for the engineering convention one should reverse all imaginary quantities.

The complex permittivity is usually a complicated function of frequency ω, since it is a superimposed description of dispersion phenomena occurring at multiple frequencies. The dielectric function $\varepsilon(\omega)$ must have poles only for frequencies with positive imaginary parts, and therefore satisfies the Kramers–Kronig relations. However, in the narrow frequency ranges that are often studied in practice, the permittivity can be approximated as frequency-independent or by model functions.

At a given frequency, the imaginary part of $\hat{\varepsilon}$ leads to absorption loss if it is positive (in the above sign convention) and gain if it is negative. More generally, the imaginary parts of the eigenvalues of the anisotropic dielectric tensor should be considered.

In the case of solids, the complex dielectric function is intimately connected to band structure. The primary quantity that characterizes the electronic structure of any crystalline material is the probability of photon absorption, which is directly related to the imaginary part of the optical dielectric function $\varepsilon(\omega)$. The optical dielectric function is given by the fundamental expression:

$$\varepsilon(\omega) = 1 + \frac{8\pi^2 e^2}{m^2} \sum_{c,v} \int W_{c,v}(E)\big(\varphi(\hbar\omega - E) - \varphi(\hbar\omega + E)\big)dx.$$

In this expression, $W_{c,v}(E)$ represents the product of the Brillouin zone-averaged transition probability at the energy E with the joint density of states, $J_{c,v}(E)$; φ is a broadening function, representing the role of scattering in smearing out the energy levels. In general, the broadening is intermediate between Lorentzian and Gaussian; for an alloy it is somewhat closer to Gaussian because of strong scattering from statistical fluctuations in the local composition on a nanometer scale.

Tensorial Permittivity

According to the Drude model of magnetized plasma, a more general expression which takes into account the interaction of the carriers with an alternating electric field at millimeter and microwave frequencies in an axially magnetized semiconductor requires the expression of the permittivity as a non-diagonal tensor.

$$\mathbf{D}(\omega) = \begin{vmatrix} \varepsilon_1 & -i\varepsilon_2 & 0 \\ i\varepsilon_2 & \varepsilon_1 & 0 \\ 0 & 0 & \varepsilon_z \end{vmatrix} \mathbf{E}(\omega)$$

If ε_2 vanishes, then the tensor is diagonal but not proportional to the identity and the medium is said to be a uniaxial medium, which has similar properties to a uniaxial crystal.

Classification of Materials

Classification of materials based on permittivity		
$\dfrac{\varepsilon_r''}{\varepsilon_r'}$	Current conduction	Field propagation
0		perfect dielectric lossless medium
$\ll 1$	low-conductivity material poor conductor	low-loss medium good dielectric
≈ 1	lossy conducting material	lossy propagation medium
$\gg 1$	high-conductivity material good conductor	high-loss medium poor dielectric
∞	perfect conductor	

Materials can be classified according to their complex-valued permittivity ε, upon comparison of its real ε' and imaginary ε'' components (or, equivalently, conductivity, σ, when accounted for in the latter). A *perfect conductor* has infinite conductivity, $\sigma = \infty$, while a *perfect dielectric* is a material that has no conductivity at all, $\sigma = 0$; this latter case, of real-valued permittivity (or complex-valued permittivity with zero imaginary component) is also associated with the name *lossless media*. Generally, when $\dfrac{\sigma}{\omega\varepsilon'} \ll 1$ we consider the material to be a *low-loss dielectric* (although not exactly lossless), whereas $\dfrac{\sigma}{\omega\varepsilon'} \gg 1$ is associated with a *good conductor*; such materials with non-negligible conductivity yield a large amount of loss that inhibit the propagation of electromagnetic waves, thus are also said to be *lossy media*. Those materials that do not fall under either limit are considered to be general media.

Lossy Medium

In the case of a lossy medium, i.e. when the conduction current is not negligible, the total current density flowing is:

$$J_{tot} = J_c + J_d = \sigma E - i\omega\varepsilon'E = -i\omega\hat{\varepsilon}E$$

where

- σ is the conductivity of the medium.

- ε' is the real part of the permittivity.

- $\hat{\varepsilon}$ is the complex permittivity.

The size of the displacement current is dependent on the frequency ω of the applied field E; there is no displacement current in a constant field.

In this formalism, the complex permittivity is defined as:

$$\hat{\varepsilon} = \varepsilon' + i\frac{\sigma}{\omega}$$

In general, the absorption of electromagnetic energy by dielectrics is covered by a few different mechanisms that influence the shape of the permittivity as a function of frequency:

- First are the relaxation effects associated with permanent and induced molecular dipoles. At low frequencies the field changes slowly enough to allow dipoles to reach equilibrium before the field has measurably changed. For frequencies at which dipole orientations cannot follow the applied field because of the viscosity of the medium, absorption of the field's energy leads to energy dissipation. The mechanism of dipoles relaxing is called dielectric relaxation and for ideal dipoles is described by classic Debye relaxation.

- Second are the resonance effects, which arise from the rotations or vibrations of atoms, ions, or electrons. These processes are observed in the neighborhood of their characteristic absorption frequencies.

The above effects often combine to cause non-linear effects within capacitors. For example, dielectric absorption refers to the inability of a capacitor that has been charged for a long time to completely discharge when briefly discharged. Although an ideal capacitor would remain at zero volts after being discharged, real capacitors will develop a small voltage, a phenomenon that is also called *soakage* or *battery action*. For some dielectrics, such as many polymer films, the resulting voltage may be less than 1–2% of the original voltage. However, it can be as much as 15–25% in the case of electrolytic capacitors or supercapacitors.

Quantum-mechanical Interpretation

In terms of quantum mechanics, permittivity is explained by atomic and molecular interactions.

At low frequencies, molecules in polar dielectrics are polarized by an applied electric field, which induces periodic rotations. For example, at the microwave frequency, the microwave field causes the periodic rotation of water molecules, sufficient to break hydrogen bonds. The field does work against the bonds and the energy is absorbed by the material as heat. This is why microwave ovens work very well for materials containing water. There are two maxima of the imaginary component (the absorptive index) of water, one at the microwave frequency, and the other at far ultraviolet (UV) frequency. Both of these resonances are at higher frequencies than the operating frequency of microwave ovens.

At moderate frequencies, the energy is too high to cause rotation, yet too low to affect electrons directly, and is absorbed in the form of resonant molecular vibrations. In water, this is where the absorptive index starts to drop sharply, and the minimum of the imaginary permittivity is at the frequency of blue light (optical regime).

At high frequencies (such as UV and above), molecules cannot relax, and the energy is purely absorbed by atoms, exciting electron energy levels. Thus, these frequencies are classified as ionizing radiation.

While carrying out a complete *ab initio* (that is, first-principles) modelling is now computationally possible, it has not been widely applied yet. Thus, a phenomenological model is accepted as being an adequate method of capturing experimental behaviors. The Debye model and the Lorentz model use a first-order and second-order (respectively) lumped system parameter linear representation (such as an RC and an LRC resonant circuit).

Measurement

The dielectric constant of a material can be found by a variety of static electrical measurements. The complex permittivity is evaluated over a wide range of frequencies by using different variants of dielectric spectroscopy, covering nearly 21 orders of magnitude from 10^{-6} to 10^{15} hertz. Also, by using cryostats and ovens, the dielectric properties of a medium can be characterized over an array of temperatures. In order to study systems for such diverse excitation fields, a number of measurement setups are used, each adequate for a special frequency range.

Various microwave measurement techniques are outlined in Chen *et al.* Typical errors for the Hakki-Coleman method employing a puck of material between conducting planes are about 0.3%.

- Low-frequency time domain measurements (10^{-6} to 10^3 Hz)

- Low-frequency frequency domain measurements (10^{-5} to 10^6 Hz)

- Reflective coaxial methods (10^6 to 10^{10} Hz)

- Transmission coaxial method (10^8 to 10^{11} Hz)

- Quasi-optical methods (10^9 to 10^{10} Hz)

- Terahertz time-domain spectroscopy (10^{11} to 10^{13} Hz)

- Fourier-transform methods (10^{11} to 10^{15} Hz)

At infrared and optical frequencies, a common technique is ellipsometry. Dual polarisation interferometry is also used to measure the complex refractive index for very thin films at optical frequencies.

Permittivity of Insulating Materials

The most important electrical parameter of insulating materials is the capacitance offered by them between given electrode systems. The capacitance thus formed is always accompanied with some losses determined by the permittivity 'ε' and the specific insulation resistancepins of the material.

The permittivity of an insulating material ε is defined as the product of absolute permittivity of free space (vacuum) 'ε_0' and the relative permittivity 'ε_r' of the material or the medium:

$$\varepsilon = \varepsilon_0 \, \varepsilon_r$$

The absolute permittivity of free space, ε_0 is constant and has a value,

$$\varepsilon_0 = 8.854.10^{-12} \, {}_{F/m}$$

On the contrary, the relative permittivity of a material ε_r is not a constant. It depends upon the thermal conditions of the material as well as the frequency and the magnitude of the applied voltage. ε_r is often mentioned as dielectric number or permittivity number in the literature. ε_r of a dielectric is defined as the quotient of the capacitances C_x to C_0, where C_x is the capacitance of a condenser constituting the given material as a dielectric and C_0 is the capacitance of the same condenser constituting vacuum as a dielectric in uniform field.

$$\varepsilon_r = \frac{C_x}{C_0} \tag{1}$$

For the mathematical analyses taking into account the polarization process in the dielectrics, it was necessary to introduce the complex relative permittivity ' $\overline{\varepsilon}_r$ ' described as:

$$\overline{\varepsilon}_r = \varepsilon_r' - j\varepsilon_r'' \tag{2}$$

where ε_r' is the real and ε_r'' the imaginary part of the complex relative permittivity $\overline{\varepsilon}_r$, which is also known as 'loss index'.

Polarization in Insulating Materials

Polarization in insulating materials is basically a phenomenon of interaction between the applied external electric field and the inherent charge carriers, the atoms, ions or molecules present in the dielectric. Not only the applied electric field gives rise to the polarization, but in turn, polarization modifies the microscopic field within the dielectric. Thus, it is a process of a reversible displacement between the positions of positive and negative charges in the molecular structure caused by the applied electric field. The interaction takes place by a force, exercised on the basic structural elements (charge carriers) of the dielectric.

The extent of polarization is analytically described by the relative permittivity ε_r. Consider a uniform field electrode system in vacuum applied an electric field E. The electric flux density $\vec{D}_0$ is given by,

$$\vec{D}_0 = \varepsilon_0 . \vec{E} \tag{3}$$

For the same electrode and the magnitude of the applied electric field, when the vacuum is replaced by an insulating material (solid, liquid or gas), the electric flux density in the dielectric $\vec{D}_{ins}$, is increased, given by,

$$\vec{D}_{ins} = \varepsilon_0 . \varepsilon_r . \vec{E} \tag{4}$$

The increase in electric flux density, '$\vec{D}_P$' is caused by polarization in the dielectric. $\vec{D}_P$, therefore, represents the material property and is given as:

$$\vec{D}_P = \vec{D}_{ins} - \vec{D}_0 = \varepsilon_0 (\varepsilon_r - 1) \vec{E}$$

$$\text{Or } \frac{\vec{D}_P}{\vec{D}_0} = (\varepsilon_r - 1) \tag{5}$$

The quotient D_p / D_0 is known as dielectric susceptibility or polarization capacity of a dielectric. It is also denoted by χ_e.

Most of the gaseous dielectrics have their ε_r nearly equal to one. Hence polarization is not an important phenomenon in gases, but in liquid and solid dielectrics it plays an important role determining the conductivity and the losses. Different types of polarization mechanisms are described under the following three main categories.

Displacement Polarization

Within a molecular bond, the positive and negative charges of individual molecule or atom are rendered to oscillate in synchronism under the influence of an applied electric field. Similar oscillations also take place between the nucleus and the electron shell of an atom, building dipoles. The displacement caused by the oscillations is proportional to the applied electric field and on removing the electric field the atoms return back to their original state. Where only this type of polarization mechanism is present, the dielectric materials are described to be 'nonpolar' and these have no dipole moment. Classic examples of such dielectrics are the gases.

Space Charge or Boundary Surface Polarization

Some heterogeneous dielectrics, for example partially crystalline insulating materials, have the positive and negative ions as free charge carriers. Without an external field, the dielectric is in neutral condition and the positive and negative charges neutralise each other. However, under the effect of an external electric field, the charge carriers in the dielectric move towards opposite polarity electrode surfaces, giving rise to a macroscopic dipole. Like the mechanism described in previous case, in this care too on removing the applied electric field, the dielectric returns to its original state, hence it is known as a type of displacement polarization.

Dielectrics having this type of polarization mechanism are also described as nonpolar, that is, these do not build a dipole without an external field. The boundary surface polarization is commonly present in heterogeneous insulating materials, such as, impregnated paper insulations, hard pressed boards, taped insulations used in electrical machines and even at voids in solid homogeneous dielectrics. Such materials normally have a low value of ε_r.

Orientation Polarization

Orientation polarization is the main polarization mechanism in liquid dielectrics. Usually, the polarization depends upon the applied electric field intensity. However, in some materials a permanent polarization is 'frozen' due to the permanent dipoles present in the dielectrics even without an external field. Such dielectrics have an asymmetry in their molecular structure and are described as 'polar'.

On applying an external electric field, the dipoles arrange themselves according to the field lines. The effect of applied external field on the dipoles is determined by their dipole moments. Depending upon the bond between the positive and the negative charges in the molecules of a material, the orientation of dipoles takes place. The effective field intensity established in the material is the quantity induced by the external source and the sources within the material itself. The orientation of dipoles may be in the form of an arrangement in an element or they may simply straighten out like a chain, which usually depends upon the local field in which the molecules are situated. The local microscopic field may not be necessarily equal to the macroscopic applied electric field.

Surface Discharge (Tracking)

PB on the surface, also known 'tracking', occur in any gaseous medium and also in vacuum at the interfaces of solid and liquid dielectrics. Surface discharges are caused due to the presence of high tangential component of electric field along a diagonal interface between any solid or liquid and a gaseous medium. Typical examples of the presence of such interfaces in practice are in bushings, insulators, cable terminations and sites in transformers. As shown in figure, when the concentric conductor of a coaxial cable is abruptly terminated, the field intensity at the point of termination becomes very high. An increased tangential component of the field is resulted in the gaseous medium which stresses it with higher field intensity. Surface discharges are caused if an interruption of the creepage current across the surface of the dielectric takes place and the field component at the location exceeds the electric strength of the medium. Surface Discharge or tracking is also a PB phenomenon.

Day in [3.9] describes surface discharge as an untidy process, whose occurrence depends upon the properties of the dielectric but their inception depends upon several other factors too. Day defines tracking as the formation of conducting path on the surface of a dielectric. In most cases, the conduction results from degradation of the dielectric or by deposition of a foreign layer on it. Hence for tracking to occur, the following are essential to be present:

- a conducting film across the surface of the dielectric

- a mechanism by which the creepage current through the conducting film is interrupted when the discharges occur.

The conducting film is usually a form of contamination, such as salt deposition (in coastal areas), carbonaceous dust from fuels and industrial or cellulose fibre deposits along with moisture. Interruption of moisture films is caused when a surface is dried due to the heating effect of the surface current. As the amount of contamination increases, so does the surface current, until the joule heating in the layer causes it to dry out locally initiating discharges between the receding edges of the wet film. On the surfaces of many dielectrics tracking may not occur under dry and clean conditions. An estimate revealed that tracking is likely only when the creepage currents of the order of 1 mA are able to develop on the surface.

The presence of a diagonal interface gives rise to high tangential and normal components of field at the surface. With the result, multiplication of charge carriers takes place right at the surface. On increasing the applied voltage, the same sequence of discharge mechanism is followed on the surface as in case of gaseous and liquid dielectrics, that is, avalanche, streamer and leader surface discharges. Since the charges get trapped over the surface; the more is the discharge magnitude, longer track lengths with numerous branches are produced depending upon the properties of the material. A typical autograph of surface discharges is shown in Fig. v taken by Nema [3.10].

The theoretical consideration for a material to have tendency to track is suggested in terms of chemical bond energies of the material by Parr and Scarisbrick [3.11]. Since the primary bonds are important when considering thermal stability of polymers, the tendency to track depends upon the proportion of bond which produces free carbon on pyrolysis, that is, by chemical decomposition of molecules as a consequence of exposure to a high temperature. It can be expressed as follows:

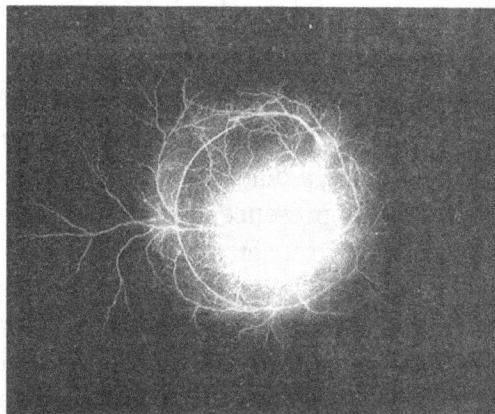

Fig. v A typical autograph of surface discharges showing stable streamer as well as leader corona, Nema [3.10].

$$\frac{T_c}{T_m} = \frac{\text{Dissociation energy of all the bonds which on breaking produce free carbon (kcal/mole)}}{\text{Total bond energy of the molecules (kcal/mole)}}$$

Smaller the fraction T_c/T_m, less is the likely hood of dielectrics to track. Experimental results revealed that if T_c/T_m is smaller than 0.3, the material has no tendency to track. Polymers which do not track are for example, polypropylene, poly-formaldehydes, etc. On the other hand, if this fraction is greater than or equal to about 0.5, the material may track easily. For example, polystyrene, phenol-formaldehydes and polycarbonates are among the polymeric dielectrics which track easily besides the ordinary glass being well known to track.

Since it is the free carbon produced by thermal decomposition which forms the conducting tracks, the degree of resistance to track of a material depends upon its chemical nature and the manner in which it breaks down when subjected to high temperatures. For example, the degradation products of PE are gaseous, therefore the polymer does not track normally except when some conducting contamination is trapped on its surface during erosion. This is the reason that in order to prevent tracking, a creepage distance is often specified to limit surface currents.

For the sake of obtaining a comparative knowledge of tendency to track of materials, IEC-112 and BS-378I have introduced an index, known as 'Comparative Tracking Index' (CTI), giving standard procedure of its measurements. The CTI can be used as a specification for the measurement of track-resistant quality of a dielectric. The CTI greater than 750 represents the upper stratum of materials in terms of track resistance. However, the use of a material having CTI higher than 750 may not guarantee that tracking will not occur under very bad contamination conditions.

Degradation of Solid Dielectrics Caused by PB

Internal discharges in solid dielectrics cause their degradation and lead to the formation of solid, liquid and gaseous products due to synthesis reactions taking place within the -electrode gap space. Organic compounds containing oxygen are readily synthesized from mixtures of CO_2, CO and H_2O undergoing discharges. Experimental investigations made by Gamez-Garcia et al. [3.12] on PE and XLPE confirmed their earlier findings that CO_2, CO, H_2O and acetophenone are the main degradation products. This was revealed by their experimental investigations of products obtained from tests on XLPE degradation due to discharge synthesis reactions, which took place within the electrode gap space. Oxalic acid and other non aromatic compounds result from discharge synthesis reactions involving CO and H_2O. Acetophenone, which evaporates from an XLPE surface after cross-linking, plays an important role in the formation of other degradation products. Benzoic acid, benzamide, toluene and other aromatic compounds result from chemical reactions of this volatile compound during the degradation process taking place in the presence of PB.

If the PB intensity is unusually high, the degradation and erosion mechanism may proceed at sufficiently fast rate. Conductive channels in different shape and size are formed within the solid dielectrics, known as ' Tree' formation. The phenomenon is known as Treeing process. It grows towards the opposite electrode with time. It may lead to complete failure of the insulating properties of the dielectric taking different amount of time which depends upon the material, the field intensity and its shape. The exact conditions determining these mechanisms are yet to be defined through more detailed investigations.

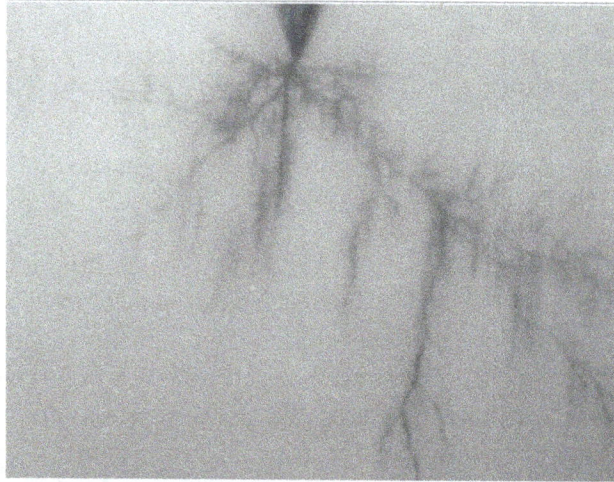

Fig. vi Development of Treeing Process in Epoxy Resin

Under one such investigation, treeing process developed in solidified resin is shown in Fig. vi. A needle electrode imbedded inside the resin block produces an extremely nonuniform field at the tip of the needle. This gives rise to PB within the solid dielectric on applying a voltage greater than U_i, the PB inception voltage. The PB causes conductive channels due to degradation of the dielectric under heat. These develop in the form of a tree extending towards the opposite electrode, the ground electrode. The development of treeing process in solid dielectrics could be caused at any localised enhancement of electric field intensity. It could be due to a void, foreign particle, metallic protrusion or ingress of water/moisture in the solid dielectric, such a formation of tree is the formation of conductive channels within a healthy material. This process reduces the life of the dielectric drastically.

From the above, it can be concluded that PB in solid dielectrics cause physical changes through chemical synthesis reactions. The nature and extent of these physical changes depend upon the particular local condition and the time factor, which play a very important role.

Breakdown and Prebreakdown Phenomena in Solid Dielectrics

The prebreakdown phenomena in solid dielectrics leading to an irreversible rupture of the material (breakdown) is extremely complicated. In this process the local electric field, heat transport, charge injection and accumulation, mechanical stress, chemical and mechanical stability are all strongly and often nonlinearly coupled. Above all, the time factor affecting the breakdown makes the prebreakdown phenomena more complicated.

Intrinsic Strength of Solid Dielectrics

As described in case of liquid dielectrics, the same definition of intrinsic strength holds good also for solid dielectrics. For a homogeneous and isotropic solid dielectric, the intrinsic strength is described as the highest value of electric strength obtainable after all known secondary effects leading to a premature breakdown (below intrinsic strength), seem to have been eliminated.

The concept remains an ideal one and its validity is always dubious, as it is very difficult to know whether an observed breakdown was or was not intrinsic. One may conclude therefore, that the observed results are more likely to be intrinsic, the more they are independent of parameters like temperature, thickness and time. It is also extremely important that the test sample is prevented from self heating and mechanical distortions caused by the field.

In order to measure intrinsic strength, preparation of the test samples is the utmost sensitive part involving a considerable amount of precision techniques. As described by Garton in [5.11], only two methods of measurement are seriously considered, the 'recessed specimen' and the 'McKeown's technique'. Lawson [4.1] measured the intrinsic strength of PE samples with the help of above mentioned specimens at different temperatures described in the following:

'Recessed Specimen'

A spherical depression is either pressed, machined or ground on a ca. 1.5 mm thick plane disc sample of the material. It nearly penetrates the specimen, leaving only a thickness of approximately 50 µm, suitable for breakdown by a reasonable voltage, as shown in Fig. vii (a). The radius of the spherical depression should be such that a uniform field is achieved. The surfaces of the recess and of the opposite plane faces are then made conducting by a technique which provides an extremely good contact with the dielectric, for example, an evaporated film of aluminum, or a graphite coating applied by repeated spraying and polished to a continuous film. Liquid electrodes have also been tried. However, limitations like inadequate cooling of the stressed area and no mechanical support for the thin apex of the specimen remain with such electrodes.

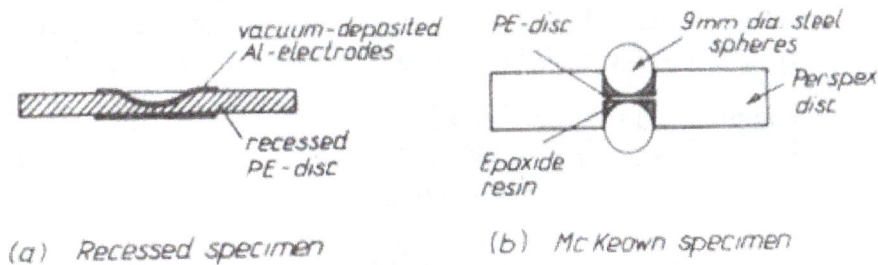

Fig. vii Specimens for the measurements of intrinsic strength, Lawson [4.1].

'McKeown Technique'

Two ball bearings used as electrodes in this specimen are located within a cylindrical hole in a disc made of suitable insulating material, such as Perspex. The lower ball is cemented in the position with an epoxyresin, and the hole filled with the same resin in liquid form. Then it is gently heated in partial vacuum until degassed. As shown in Fig. viii (b), the polyethylene test sample in the form of a thin film disc is placed in the hole on the top of the ball and the liquid is degassed again. Before encapsulation, the polyethylene film sample were treated chemically by immersion in concentrated chromic acid for five minutes to oxidize their surfaces and make a bond with the epoxide resin. The top ball is then inserted and the whole specimen cured until solid conditions are achieved. The thickness of the test sample assembled can be obtained by measuring the total thickness outside the balls and subtracting twice the diameter of the ball. A proper adhesion of the test sample material with the epoxyresin must be achieved to avoid any partial discharges or tracking on the interfaces. The specimen thus prepared may further require a chemical treatment with a solution of sodium in liquid ammonia in order to prevent corona discharges on the surfaces.

As reported by Garton, both these techniques have proved extremely satisfactory at room temperatures for measuring the intrinsic strength of low-loss polymers. However, the results obtained by McKeown specimens have been consistantly higher than those obtained by the 'recessed' specimens.

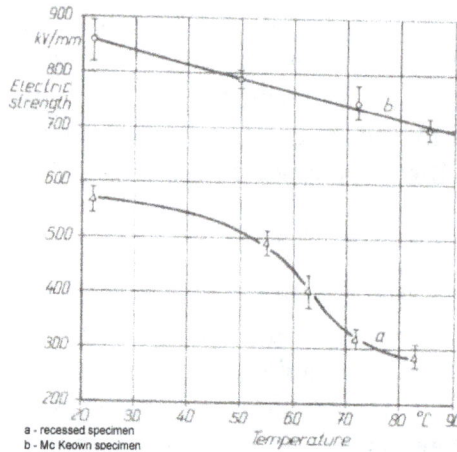

Fig. viii Variation of intrinsic electric strength of PE specimens with temperature, Lawson [4.1] .

Lawson [4.1] used McKeown techniques to measure the intrinsic strength of PE with increasing temperature between 20-85°C, as shown in Fig. viii. In his experiments, the breakdown tests were carried out under transformer oil to prevent flashover due to surface discharge. The applied voltage raised linearly with time, was obtained from a smoothed half-wave rectifier source. In these experiments, the breakdown occurred in about 15-20 s. In all cases the number of specimens tested were from six to eight, and 95% confidence limit of the mean was considered.

Compared to the results obtained from recessed specimens, the fall of intrinsic strength with increasing temperature was greatly reduced by using McKeown specimens. Besides, altogether higher values of electric strength, of the order of 800 kV/mm, were obtained for PE.

The theories developed for intrinsic breakdown aim to arrive at a condition for an irreversible instability of the dielectric which is electronic in nature. These simply assume that in case of intrinsic breakdown the irreversible electronic catastrophe produces conditions in which the lattice is destroyed which is a temperature dependent phenomenon, Lawson [4.1].

The general aim of the measurement of intrinsic electric strength of the dielectrics has been to reveal the extreme complexity of the breakdown mechanism at very high electric fields rather than to find out electric strength. A breakdown in real life always occurs through some secondary mechanism or an external cause. Therefore, to conclude that a breakdown was intrinsic is mostly dubious. In practice, the solid dielectrics are stressed with maximum electric stress not greater than 3% of such high values of intrinsic strengths.

Thermal Breakdown

Since it develops very rapidly, the intrinsic breakdown is assumed to have electronic origin. The most simple theoretical approach for intrinsic breakdown has been to derive a theory for an electronic instability in the dielectric subjected to a uniform field, O'Dwyer [4.2]. While measuring the intrinsic strength, if there exists a cause which leads to other than electronic instability, the value of electric strength measured may be anywhere below intrinsic strength. One such cause is 'thermal instability', particularly in the dielectrics having sufficient conductivity. The conductivity is found to increase with electric stress, but the rate of increase may not be linear, Lawson [4.1]. The 'hot electron' theory of breakdown is solid dielectrics developed by Frohlich and Paranjape [4.2] also predicted a stress-enhanced conductivity, which may cause thermal failure before a true instability (intrinsic failure) occurs. This theory involves lattice temperature rise processes and is derived for intrinsic or near intrinsic breakdowns.

The experimental concept of 'thermal breakdown' is different as it is based upon self heating of the dielectric due to power losses. Because of power loss due to conductivity, polarization or other forms of dielectric losses, heat is produced continuously in electrically stressed dielectrics. Depending upon the magnitude of the applied voltage, its period of application and the conduction of heat the dielectric temperature rises. If the heat generated within a dielectric system equals the dissipation of heat to the surroundings by thermal conduction, the temperature rises to an equilibrium value, and a thermally balanced and stable operation of the insulation system takes place. But in practice, the power losses in a dielectric are some function of temperature also. Since the ionic conductivity increases with increasing temperature, the power losses increase with temperature. If the dissipation of heat by cooling processes is not adequate, it is possible that an unstable state may arise in which the temperature increases without limit causing a breakdown.

Mechanism of Breakdown in Extremely Non-uniform Fields

While designing electrode configurations in practice, it is tried to avoid extremely non-uniform fields, especially for solid insulations. Thus, one may say that solid dielectrics are normally not stressed with extremely sharp non-uniform fields in practice. But, the pre breakdown process generally begins at an extreme intensity or distortion of local fields. Under the action of this high local electric field, a solid dielectric may either globally or locally loose its electrical insulating property and mechanical stability, leading to its complete or partial disintegration. A review of the breakdown and pre breakdown phenomena in solid dielectrics made by Zeller [4.3] has brought out certain important conclusions.

At fields closer to local breakdown i.e. PB inception at the tip of a sharp electrode, massive space charge injection from the electrodes into the dielectric is inevitable. The local electric stress dominated by the space charge may increase considerably. The electromechanical action due to electric force causes dielectric instabilities in solids, giving rise to initiation and growth of electrical trees, a conductive path with the dielectric.

Treeing a Prebreakdown Phenomenon in Polymeric Dielectrics

Treeing is an electrical pre-breakdown phenomenon. The name is given to the type of damage caused to the solid dielectrics, which progresses through its certain part under electric stress in extremely non-uniform fields. The path of its growth appears like a miniaturized tree, hence the name given to this phenomenon. Treeing process normally begins at a very sharp, pointed electrode tip due to internal partial breakdown which take placeuner extremely nonuniform field conditions. Extreme field distortions at protrusions in the dielectrics may also give rise to the development of treeing process.

Treeing may occur and develop slowly due to PB. In the presence of moisture it may develop slowly even without any measurable PB. However, treeing develops rapidly when very high impulse voltage is applied. While treeing is generally associated with ac and impulse voltages, it also takes place under high dc stresses. Evidences are there that the treeing process is aggravated by the presence of moisture, chemical environment, voids and contaminants such as foreign particles. Treeing may or may not be followed by complete electrical breakdown of the part of the dielectric where it occurs. However, in solid polymeric dielectrics it is one of the most common cause of electrical failure. The whole process may or may not take long time for it strongly depends upon the local conditions.

Forms of Treeing Patterns

The great variety of patterns which appear like stems and branches comprising an electrical tree, are described in the literature with large number of corresponding names of the figures such as, dendrite, fan, plume, delta, bush, broccoli, string and bow tie, etc. Eichhorn [4.4]. The reader should therefore not get confused as they all represent the same phenomenon of treeing under different local conditions. In Fig. ix (a), a tree, and in ix (b) a bush having tree like branches are shown. These were produced in Piacryl, a transparent solid dielectric, in extremely nonuniform field between needle and plane by Pilling [4.5].

Besides depending upon the in homogeneity of the dielectric material, the pattern of growth of treeing process also depends upon the radius of the tip of needle, diameter of needle, gap length 'd' as well as shape and size of the plane electrode, Loffelmacher [4.6].

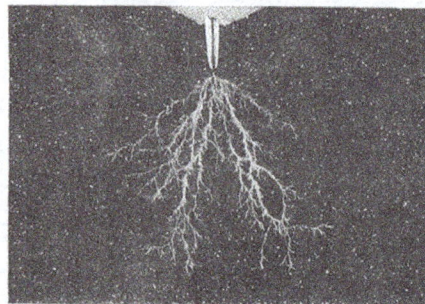

(a) A Tree in Piacryl

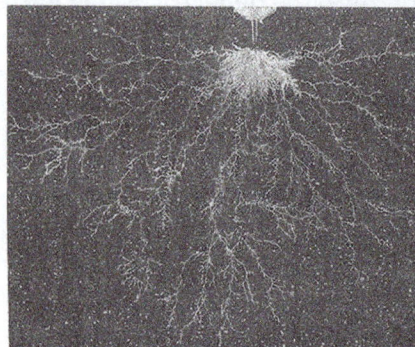

(b) A bush like tree in Piacryl

Fig. ix Treeing in Piacryl, Pilling [4.5].

In figure x , bow tie type tree is shown, Eichhorn [4.4]. Such a tree developed at a void on the surface of the dielectric and grew up along it on both sides.

⊢————⊣ *0,1 mm*

Fig. x A bow tie type tree

References

- Peter Y. Yu; Manuel Cardona (2001). Fundamentals of Semiconductors: Physics and Materials Properties. Berlin: Springer. p. 261. ISBN 3-540-25470-6

- John H. Moore; Nicholas D. Spencer (2001). Encyclopedia of chemical physics and physical chemistry. Taylor and Francis. p. 105. ISBN 0-7503-0798-6

- Murphy, E. J.; Morgan, S. O. "The Dielectric Properties of Insulating Materials" (PDF). Retrieved October 2, 2012

- Hartmut Haug; Stephan W. Koch (1994). Quantum Theory of the Optical and Electronic Properties of Semiconductors. World Scientific. p. 196. ISBN 981-02-1864-8

- Linfeng Chen; V. V. Varadan; C. K. Ong; Chye Poh Neo (2004). "Microwave theory and techniques for materials characterization". Microwave electronics. Wiley. p. 37. ISBN 0-470-84492-2

High Voltage Measurement Techniques

In order to measure different types of voltages there are laboratory techniques. The different types of voltages are ac power frequency, dc voltages and impulse voltages. The chapter closely examines high voltage measurement to provide an extensive understanding of the subject.

High Voltage Test

High test voltages generated in laboratories require special techniques for their measurement. Each type of high voltage namely; ac ($\sim$), dc (=) and impulse ($\wedge$) can be suitably measured with more than one different techniques. The choice of the method of measurement in a laboratory is a matter of convenience, cost and the accuracy required for the measurement besides the safety. The complications in measurements increase with the magnitude of the voltage desired to be measured. The problems confine mainly with respect to the large structures of the device and the electric stress control techniques applied to prevent PB and flash overs.

The high voltage measurement techniques in electric power systems are different. These are not discussed in this course.

Different measurement techniques adopted for the three different types of voltages are given in the table.

As it can be seen in this table, the earliest technique of sphere gap and the modern and most accurate technique of potential divider are applicable for the measurement of all the three types of voltages and both their polarities. The sphere gap technique, used widely for over a century, is not used any more. The disadvantages of this method are many. It does not provide a continuous measurement of the voltage and a flashover between the two spheres is essential. This measurement technique is not an accurate one. It may give error upto 3% in the measurement.

The electrostatic voltmeter, suitable for the measurement of ac (rms) and dc voltages, is one of the most accurate instrument. However, due to its size, cost and adjustment problems, it is not used very widely. For the measurement of ac power frequency voltage, the simplest way is to measure the voltage on the primary side or at the low voltage side of the ac test transformer and calibrate the dial of the low voltage voltmeter by multiplication of the turns ratio to get the high output voltage. However, this technique could give rise error in two ways. First, in case of high capacitive loads, the leading current output from the test transformer results in higher actual voltage output than measured with the help of trunks ratio at the primary side, the well known Ferranti effect as shown in Fig. i. Secondly, if the output voltage in not perfectly sinusoidal, the peak value is not $\sqrt{2}$ times the rms value. To overcome this problem, measurement of peak value of the output voltage at the high voltage side is recommended.

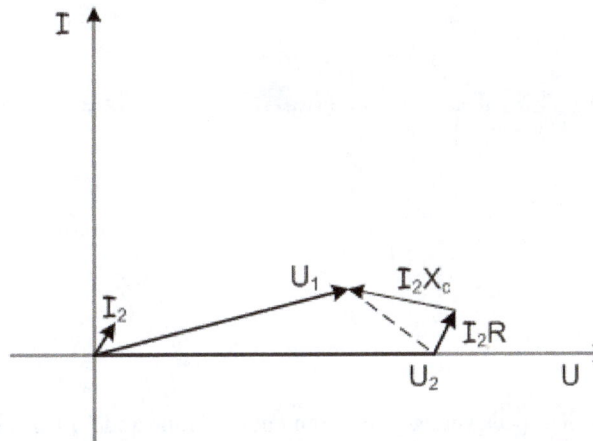

Fig. i Phaser diagram of the output voltage with leading (capacitive) current

Table: High Test Voltage Measurement Techniques

Type of Voltage	Method or Technique
ac power frequency	(i) Primary or low voltage measurement (rms)
	(ii) Sphere gap (peak)
	(iii) Electrostatic voltmeter (rms)
	(iv) Series impedance and ammeter
	(v) Potential dividers and oscilloscope (resistive, capacitive or mixed)
	(vi) Peak voltmeters with dividers
dc voltages	(i) Sphere gap, either polarity
	(ii) Series resistance and microammeter
	(iii) Resistive potential divider
Impulse voltages (li & si) either polarity and other high frequency high voltages	(i) Sphere gap
	(ii) potential dividers, oscilloscope (capacitive, resistive & capacitive)
	(iii) Peak voltmeters with series capacitors and potential dividers

Global Breakdown or rupture of insulation properties of dielectrics in general take place at the peak of the voltage applied. The power frequency ac voltage may not always have perfect sinusoidal waveform, the dc output voltage may have ripple and for the impulse voltage the peak is significant. Hence, the measurement of peak magnitude of these voltages is therefore of great importance.

Chubb-Fortescue Method

The most simple yet accurate method for the measurement of peak values of ac power frequency voltage was proposed by Chubb and Fortescue in 1913. The circuit diagram for this method is shown in Fig. ii.

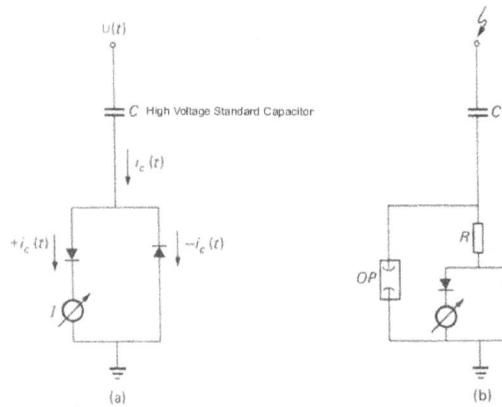

Fig. ii ac peak voltage measurement by Chubb and Fortescue.
(a) Fundamental circuit. (b) Recommended, actual circuit.

The measuring instrument in this circuit, an ammeter, reads the magnitude of charge per cycle or the mean value of the current

$$i_c(t) = C\frac{dU}{dt}$$

hence,

$$1 = \frac{1}{T}\int_{t_1}^{t_2} i_c(t)dt = \frac{C}{T}\int_{t_1}^{t_2} dU = \frac{C}{T}(|U_{+max}| + |U_{-max}|) \qquad (1)$$

the limits of integration are shown in Fig. iii.

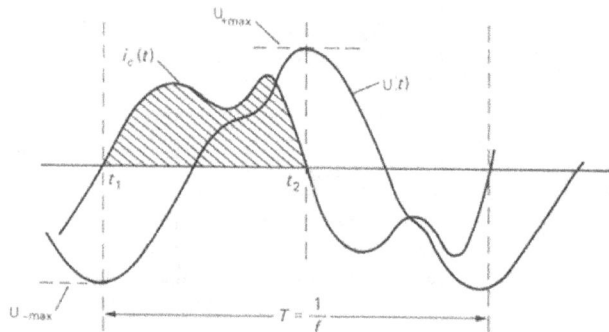

Fig. iii Diagram of voltage U(t) and current $i_c(t)$ from circuit Fig. 30.2(a)

If both, the +ve and -ve, peak values are equal, the expression for the charging current of the HV capacitor can be

$$1 = CfU_{P-P} = 2CfV_{max} \qquad (2)$$

With knowledge of exact value of f and C, the value of dc current measured gives $2U_{max}$, the actual peak to peak value of the ac voltage.

Peak Value Measurement

A number of rectifier circuits have been proposed for the measurement of peak value of ac power frequency and impulse voltages to work with capacitive voltage dividers. The first circuit of its kind was proposed by Davis, Bowdler and Strandring in 1930, shown in Fig. iv.

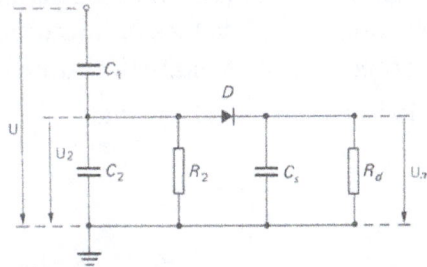

Fig . iv (a) Simple crest voltmeter for ac measurement, according to Davis, Bowdler and Standring

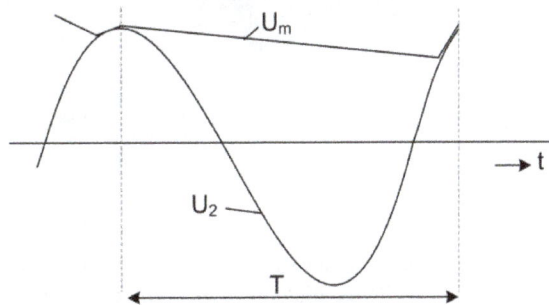

Fig. iv (b)

In this circuit, the measuring capacitors C_m is charged to the peak value , the lower arm voltage of the capacitive voltage divider. If U_2 is decreased, C_m will hold the charge and the voltage across it will not change. Because of the reverse bias current of the diode, there is a danger that U_2 will gain negative dc component over the time. Hence Um will no longer follow the change in U_2 . Therefore, discharge resistors R_2 and R_d must be included in this circuit. The time constant $R_d C_m$ should be between 0.5 - 1 sec to obtain a good response. When the circuit response conditions are taken care, the peak value of the voltage to be measured is given by the relation

$$\hat{U}_1 = \frac{C_1 + C_2}{C_1} \hat{U}_m \qquad (3)$$

Sphere Gap

This is one of the oldest technique adopted for the measurement of all the types (dc =, ac ~ and impulse) high voltages of either polarity. It remained the most widely used method for decades. The field between two identical spheres is a classical example of "weakly nonuniform" field. The breakdown characterstic of such a gap is linear for the gap distances not greater than the radius of the spheres. Measurement voltage is made as a function of minimum distance at which it can flash over or sparkover.

The breakdown voltage of the gap does not depend upon the duration of application of voltage and also not upon its variation with time.

Construction of the Sphere Gap

A sphere gap consists of two adjacent metal spheres of equal diameters whose separation gap distance is variable. The ability to respond to peak values of impulse λ voltages, if the time to peak is not too short ($\gtrsim$ 1-3 μsec), is governed by a short statistical time lag or the delay in time required for the availabiilty of an electron to initiate an electron avalanche and hence the breakdown.

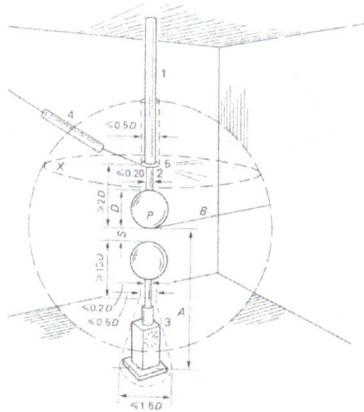

Fig. v Vertical sphere gap. 1. Insulating support. 2. Sphere shank. 3. Opening gear, showing maximum dimensions. 4. High-voltage connection with series resistor. 5. Stress distributor, showing maximum dimensions. P. Sparking point of .v. sphere. A. Height of P above ground plane. B. Radius of space free from external structures. X. Item 4 not to pass throuhg this plane within a distance B from P. Note: This figure is drwan to scale for a 100-cm sphere for gap spacing equal to their radius.

- Within the limited gap distance between the spheres, a weakly nonuniform field exists, hence no PB or corona appear before the complete breakdown or flashover.

- For a given dimension of the diameter/radius of the spheres and a gap distance between them, the magnitude of the voltage at which the breakdown occured is read from the standard calibrated tables available for different types of voltages and for both their polarities.

- The inaccuracy of measurement of voltage by this method is about 3%.

Sphere gap in an HV laboratory

- The accuracy to measurement of voltage with sphere gap is considerably affected by earthed objects around the gap since they form stray capacitance.

Correction Factor

- Since in general the actual air density during measurements differ from the reference conditions, the breakdown voltage of the air gap is given as

$$U_b = k_d\, U_{bo}$$

where U_{bo} corresponds to the table values and k_d is a correction factor related to air density.

- The actual relative air density (RAD) is given in general terms by

$$\delta = \frac{p}{p_0}\frac{273+t_0}{273+t} = \frac{p}{p_0}\cdot\frac{T_0}{T}$$

where

p_0 = air pressure for standard conditions

p = air pressure at the test conditions

$t_0 = 20\ ^{\circ}C$

t = temperature in $^{\circ}C$ at the test conditions.

- The correction factor k_d, given in table below for increasing values of RAD are slightly non linearly. However, it can be seen from the table that the correction factor k_d is almost equal to RAD, δ.

Table: Air density correction factor

Relative air density RAD, δ	Correction factor k_d
0.70	0.72
0.75	0.77
0.80	0.82
0.85	0.86
0.90	0.91
0.95	0.95
1.00	1.00
1.05	1.05
1.10	1.09
1.15	1.13

Disadvantages:

- For the measurement, breakdown in the gap has to take place.
- Continuous measurement of the voltage is not possible.
- This method is not very accurate.

Electrostatic Voltmeter

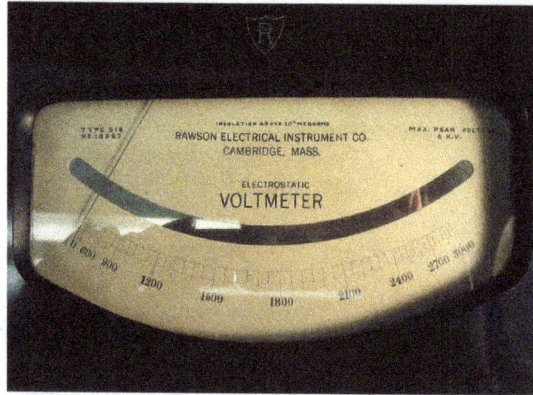

Electrostatic voltmeter

Electrostatic voltmeter can refer to an electrostatic charge meter, known also as surface DC voltmeter, or to a voltmeter to measure large electrical potentials, traditionally called electrostatic voltmeter.

Charge Meter

A surface DC voltmeter is an instrument that measures voltage with no electric charge transfer.

It can accurately measure surface potential (voltage) on materials without making physical contact and so there is no electrostatic charge transfer or loading of the voltage source.

Explanation

Many voltage measurements cannot be made using conventional contacting voltmeters because they require charge transfer to the voltmeter, thus causing loading and modification of the source voltage. For example, when measuring voltage distribution on a dielectric surface, any measurement technique that requires charge transfer, no matter how small, will modify or destroy the actual data.

Principle of Operation

In practice, an electrostatic charge monitoring probe is placed close (1 mm to 5 mm) to the surface to be measured and the probe body is driven to the same potential as the measured unknown by an electronic circuit. This achieves a high accuracy measurement that is virtually insensitive to variations in probe-to-surface distances. The technique also prevents arc-over between the probe and measured surface when measuring high voltages.

Voltmeter

The operating principle of an electrostatic voltmeter is similar to that of an electrometer, it is, however, designed to measure high potential differences; typically from a few hundred to many thousands volts.

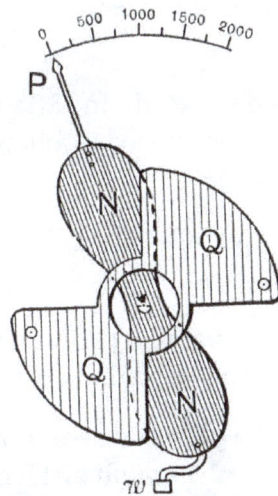

Electrostatic voltmeter operation

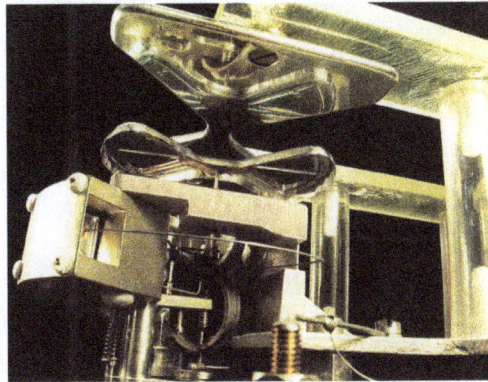

Electrostatic voltmeter mechanism

Principle of Operation

Electrostatic voltmeter utilizes the attraction force between two charged surfaces to create a deflection of a pointer directly calibrated in volts. Since the attraction force is the same regardless of the polarity of the charged surfaces (as long as the charge is opposite), the electrostatic voltmeter can measure both direct current and alternating current.

Typical construction is shown in the engraving. The pivoted sector NN is attracted to the fixed sector QQ. The moving sector indicating the voltage by the pointer P and is counterbalanced by the small weight w. In newer instruments the weight is replaced by a spring, thus allowing the meter to be used both in horizontal and vertical positions. This form of design is shown in the photograph of the mechanism. The fixed sector is insulated from the rest of the meter. The butterfly shaped moving sector, made out of a thin aluminum foil, is pivoted below. Both the fixed and the moving sectors are highly polished and without any sharp corners to minimize high electrical stress areas. The movement of the sector is damped by the air vane attached by a curved piece of wire.

- Electrostatic Voltmeters produced upto 1000 kV rated voltage are suitable for the measurement of ac power frequency and also for higher frequency rms voltages. They can also measure dc voltages.

- For higher voltage range compact voltmeters with SF_6 gas or vaccum insulation gap are also produced.

- These voltmeters were suggested by Kelvin in 1884 for the measurement of rms value of power frequency voltage. They are developed to follow the Coulomb's law which defines the static electric field as a field of force.

- The field produced between two parallel plate electrodes with shaped profile brims, is a uniform field.

- If the voltage applied across these parallel plates having a gap distance 'd' is U, then the uniform field produced in the gap between them will have an intensity E equal to U/d.

- The attracting force F between the plates on area A of the electrodes is equal to the rate of change of stored electrical energy W_{el} per unit distance in the capacitance formed between the plates.

Therefore

$$F = -\frac{\partial W_{el}}{\partial d}$$

Or

$$|F| = \frac{dW_{el}}{dd} = \frac{d}{dd}(\frac{1}{2}CU^2)$$

$$= \frac{1}{2}U^2\frac{dC}{dd} = \frac{1}{2}U^2\varepsilon_0\frac{d}{dd}(\frac{A}{d}) \text{ (since } \varepsilon_r=1)$$

$$= \frac{1}{2}\varepsilon_0 U^2 \cdot \frac{A}{d^2}$$

Or

$$|F| = \frac{1}{2}\varepsilon_0 AE^2$$

Where ε = permittivity of the insulating medium

d = gap distance between the parallel plate electrodes of area A.

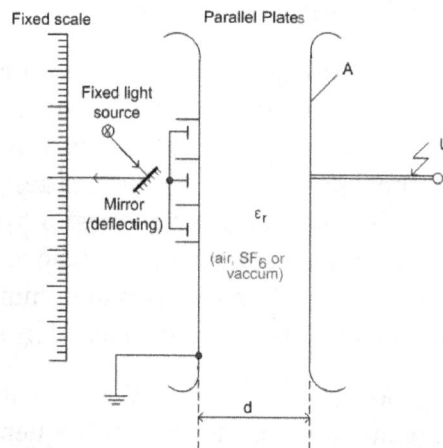

Fig. vi Schematic of an Electrostatic Voltmeter

- The attracting force is always positive independent of the polarity of the voltage. If the voltage is not constant, the force is also time (frequency) dependent. Then the mean value of the force is used to measure the rms value of the voltage. Thus

$$\frac{1}{T}\int_0^T F(t)dt = \frac{\varepsilon_0 A}{2d^2}, \frac{1}{T}\int_0^T u^2(t)dt = \frac{\varepsilon_0 A}{2d^2}(U_{rms})^2$$

where T is a proper integration time.

- Electrostatic voltmeters are arranged such that one of the electrodes or a part of it is allowed to move.

- Thus electrostatic voltmeters are rms indicating instruments if the force integration and its display follows the abvoe equation.

- Various voltmeters developed differ in their use of different methods of restoring forces required to balance the electrostatic attraction. This can be achieved by suspension of the moving electrode on one arm of a balance or its suspension on a spring or the use of a pendulous or torsional suspension.

- The small movement is generally transmitted and amplified by a spot light and mirror system, but many other systems have also been used.

- The electrostatic measuring device can be used for absolute voltage measurements since the calibration can be made in terms of the fundamental quantities of the gap length and forces.

- For a constant electrode separation 'd' the integrated forces increase with $(U_{rms})^2$ and thus the sensitivity of the system for low ranges of the rated voltages of the instrument is small. This disadvantage is overcome, however by varying the gap length in appropriate steps.

- The high pressure gas or even high vacuum between the electrodes provide very high resistivity, therefore the low active power loss.

- The measurement of voltages lower than about 50V is , however not possible as the forces become too small.

- The load inductance and the electrode system capacitance, however, form a series resonant circuit which must be damped, thus limiting the frequency range.

- Sectional view of a 1000 kV standard electrostatic voltmeter using SF_6 gas is shown below in Fig. vii.

- A five section capacitor and oil insulated bushing is used to bring the extremely high voltage into the instrument metal tank, filled with pressurized SF_6

- The HV electrode and earthed plane provide uniform electric fields in the region of the 5 cm diameter disc set in a 65 cm diameter guard plane.

- The measurement accuracy of this instrument is of the order of 0.1%.

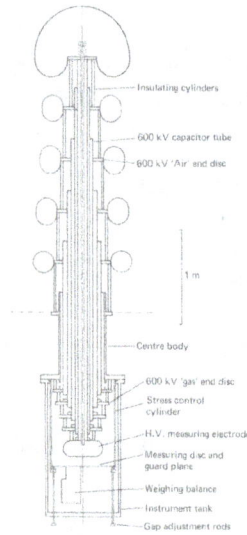

Fig. vii Sectional view of 1000-kV electrostatic voltmeter

Principle of Potential Dividers

For the measurement of High Voltages of any type, the best technique is by dividing the voltage. The HV potential divider arms could be pure capacitive, pure resistive or a combination of the two. The essential requirement is that the wave shape to be measured is correctly reproduced on the oscillograph with a known voltage reduction ratio. The general potential divider technique is shown in the following Figure viii.

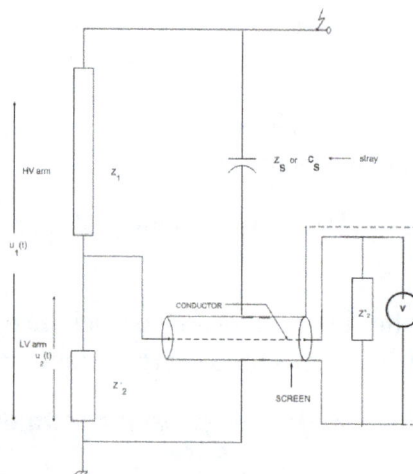

Fig. viii Voltage Divider principle

$$\frac{U_1}{U_2} = \frac{Z_1 + Z_2}{Z_2} \approx \frac{Z_1}{Z_2}$$

Z_1 - the high voltage arm of the divider. It could be pure C, pure R or combination of R and C as per requirement.

Values of Z_1 and Z_2 are chosen such that the potential across Z_2 is around 100 V.

- The voltage division ratio can be given as;

$$\frac{Z_1}{Z_2} \gg 1$$

For the circuit in Fig. viii $Z_2 = \left(\frac{1}{Z_2'} + \frac{1}{Z_2''} \right)^{-1}$

where

Z_2'' = terminal impedence of oscilloscope

For the measurement of high test voltages the technique of potential dividers is most accurate. It provides a continuous measurement of all the types of applied high voltages i.e. dc, ac and impulse. However, the suitability of the type of divider depends upon not only the type of voltage but also upon the range of magnitude of the voltage to be measured.

The types of potential dividers or basically the construction of the HV arm and their suitability are described in the following:

1. Pure Capacitance Voltage Dividers:

The HV arm of the capacitive voltage dividers, usually having their capacitance in pF range, are compressed gas or vacuum capacitors. The nF range HV capacitors are built in stack or in series a number of capacitors made out of polypropylene or paper filled with oil. In the absence of stray capacitances to earth with such HV capacitors, these provide desired exact value of low capacitance and small dimensions of the HV capacitive arm. The value of low voltage arm of the divider is normally chosen in µF range. Compressed gas (SF_6) filled HV capacitors are built even above 1000 kV (rms) rating. These can also be used as standard capacitors. In Fig. ix, a photograph and a sketch of such a capacitor are shown. The value of inductance of the long lead (in mts) connecting the low voltage electrode inside such a capacitor causes problems in giving a good response of the divider. Necessary constructional precautions are therefore taken in their design.

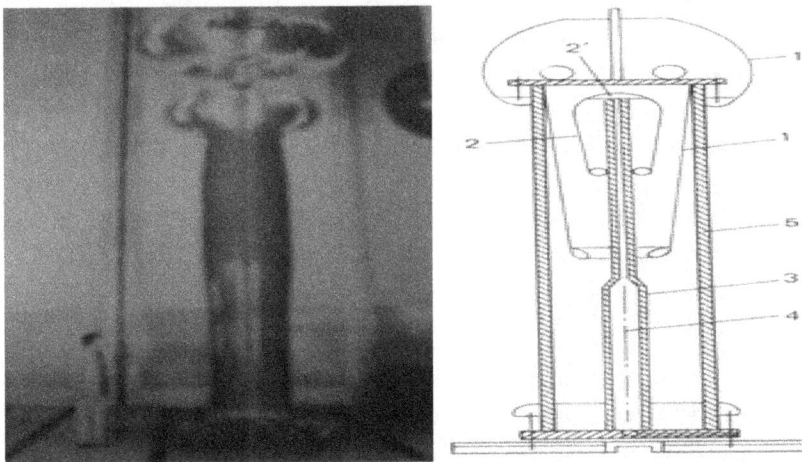

Fig. ix (a) Standard (compressed gas) capacitor for 1000 kV rms (Micafil, Switzerland) (b) Cross-section of a compressed gas capacitor. 1. Internal HV electrode. 1'. External HV electrode. 2. Low-voltage electrode with guarding, 3. Supporting tube. 4. Coaxial connection to l.v. sensing electrode. 5. Insulating cylinder.

Pure capacitance voltage dividers are suitable for the measurement of power frequency and impulse high voltages. However, the cost of the HV capacitors is high and their cost and size increases with their voltage rating.

2. Pure Resistance Potential Dividers

The high potential drop across a resistive arm by simple Ohm's law is used to reduce the voltage to a measurable quantities, circuits for such measurement are shown in Fig. x (a) and (b)

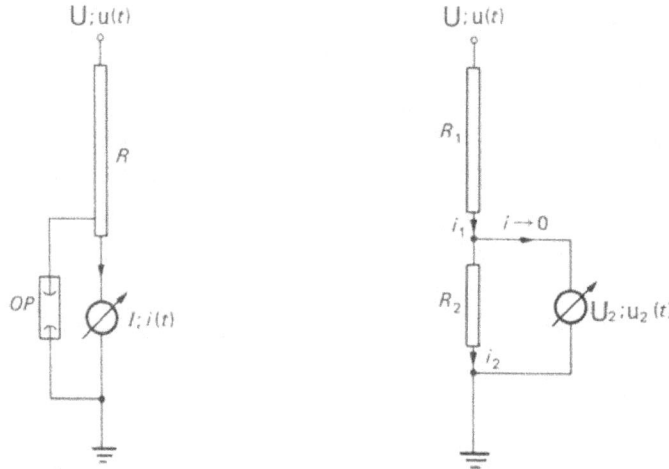

Fig. x Measurement of high dc and ac voltages by means of (a) ammeter in series with resistor R; (b) voltage divider R1, R2 and voltmeter of negligible current input. OP, Over-voltages protection.

Such pure resistive arm or divider can be used for the measurement of dc, ac as well as impulse voltages. The wave forms of ac and impulse voltages could also be recorded with the help of an oscilloscope. However, while measuring very high voltages, the resistive arm inductance and stray capacitance cause large error in the accuracy of measurement and hence in response of the measuring device.

The wire-wound resistors used earlier were made of constantan, Cu & Ni, manganin, cu, Mn, Ni or Gr and also of nichrome. To overcome the error caused due to inductance of the wire wound resistors, thin film resistors are now used. A conducting element thin film is deposited on the surface of a glass or ceramic rod or tube. The value of the desired resistance is obtained by varying the thickness of the thin film. Such films could be of carbon, metal oxide; stannic and antimony oxides. Nickel-Chromium thin film is obtained by their evaporation in vacuum. Resistive voltage dividers are produced both screened (also known as shielded) and unscreened. Since the effect of stray capacitance is not uniform, it causes a non-uniform potential gradient along its length.

3. Mixed Resistive and Capacitive Dividers

The problem caused by the stray capacitance in resistive voltage divider is over come by placing actual capacitors in parallel with the resistor units of the HV arm. This gives rise to a mixed resistive and capacitive divider used for highest voltage range of impulse voltage measurement. These are also known as 'damped capacitive divider' or RC divider. These were first introduced by Elsner in 1939, [6.1] . The simplified equivalent circuit of such a divider is shown in Fig. xi.

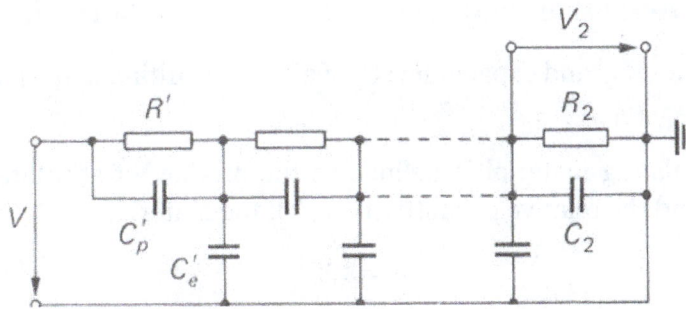

Fig. xi Simplified equivalent circuit for parallel-mixed resistor-capacitor dividers

In this Figure R' represents the resistance of single element of the HV arm where a number of them are in series. C'_p represents the capacitance across each of such element and C'_e, is the stray capacitance to earth of the element. A photograph of such a 4.5 MV voltage divider is shown in Fig.xii.

Fig. xii Damped capacitor voltage divider for 4.5-MV impulse voltage, outdoor

Non Destructive Testing of Electrical Equipment

It implies assessment of the quality of electrical insulation finish provided to the equipment. It is required in order to ensure satisfactory service of the equipment over the stipulated life span. The measurement techniques adopted are mainly electrical. As it implies, the non-destructive test measurements should not cause any damage to the equipment yet reveal the quality and condition of the dielectric performance.

Any dielectric between two electrodes forms capacitance. The most important electrical parameter of insulating materials is the capacitance offered by them between the given electrode system. The capacitance thus formed is always accompanied with losses determined by the relative permittivity ' ε_r ', the specific insulation resistance " ρ_{ins} " k and the dielectric loss factor 'tan δ ' of the material. These represent the insulating properties of the dielectric and depend upon the temperature, frequency and magnitude of the applied voltage, beside the composition of the dielectric itself.

Measurement of Dielectric Properties of Insulating Materials

The properties $\tan \delta$, ε, R_{dc} and capacitance C_x of a given insulting material are explained in the following:

Permittivity of an insulating material is defined as the product of absolute permittivity of free space (vacuum) 'ε_0' and the relative permittivity 'ε_r' of the material.

$$\varepsilon = \varepsilon_0 \varepsilon_r$$

whereas 'ε_0' is a constant and has a value 8.854×10^{-12} F/m, the relative permittivity, 'ε_r' also known as permittivity number, is defined as:

$$\varepsilon_r = \frac{C_x}{C_0}$$

Where C_x is the capacitance of a condenser constituting the given material as dielectric C_0 and is the capacitance of the same condenser constituting vacuum in place of the dielectric.

ε_r is a function of temperature of the dielectric. It also varies with the frequency and magnitude of the applied voltage. Hence ε is not a constant.

Insulation Resistance 'R_{dc}' of a dielectric is represented by the dc resistance offered by the material. It is generally described as 'specific insulating resistance 'P_{ins}' which is reciprocal of the dc conductivity 'K_{dc}'

$$\rho_{ins} = \frac{1}{k_{dc}} \ \Omega m$$

Considering a block of an insulating material in uniform field between two plates of area A at a distance d apart and applied a dc voltage U shown below.

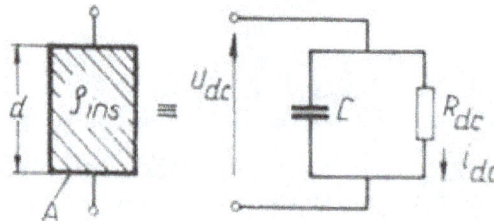

Fig. xiii An insulating material and its equivalent circuit diagram

The insulation resistance offered by this block of dielectric can be written as:

$$R_{dc} = \rho_{ins} \frac{d}{A}$$

$$I_{dc} = \frac{U}{R_{dc}} = \frac{U.A}{\rho_{ins}.d} = \frac{E.A}{\rho_{ins}} = k_{dc}.A.E$$

(Since E = U/d for uniform field)

The dc conductivity of a dielectric is a function of temperature and time for which the voltage is applied. It initially depends upon the orientation of dipoles in the material. Then by the movement of free charge carriers and ultimately by the development of space charge infront of the electrodes. R_{dc} can be directly measured with the help of a Megaohm meter.

A schematic illustration of the variation k_{dc} of with respect to the time of application of direct voltage for an insulating oil is shown below:

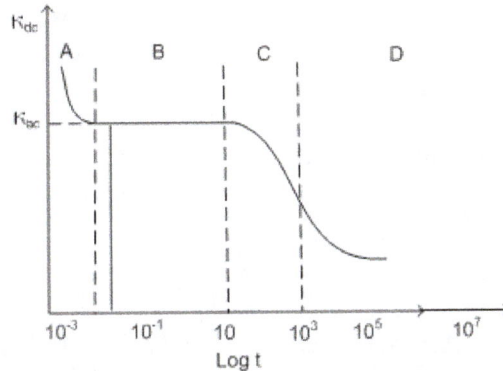

Fig. xiv A Schematic of dc conductivity of an insulating oil with respect to the time of applied voltage

In every dielectric matter, including those which have a low concentration of free charge carriers, a conductive current through the dielectric is always present on applying a voltage across the volume of the dielectric.

Besides, a small surface current determining the surface resistivity also may flow.

Accordingly two different conductivities, volume and the surface conductivities of a dielectric, are distinguished.

Reciprocal of these conductivities give volume and surface resistivities respectively.

- Specific insulation resistance ρ_{ins} , represents the specific volume resistance 'ρ_v' or the volume resistivity of the dielectric. It has a unit of Ωm.

- Specific surface resistance 'ρ_s' or the surface resistivity has accordingly a unit of Ω only.

When an alternating voltage is applied to dielectrics, they undergo active as well as reactive power losses in the capacitance formed by the dielectric. The dielectric losses vary not only with the magnitude of the applied voltage but also with the frequency of the voltage and the temperature of the dielectric. The magnitude of the losses are best assessed with the knowledge of "Dielectric Loss Tangent, $\tan\delta$". Like the relative permittivity or the permittivity number, ε_r, $\tan\delta$ is also a function of

$$\tan\delta = f(v, f, v)$$

Dielectric Loss Tangent

'$\tan\delta$' is defined as the quotient of active to reactive power loss in a dielectric when an alternating voltage is applied.

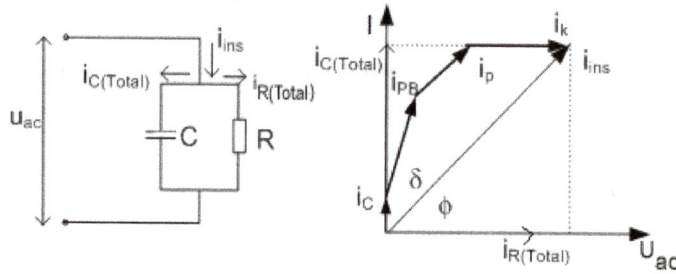

Fig. xv Conductive mechanisms in insulating materials with alternating voltages

where

i_c = capacitive charging current

i_{PB} = partial discharge impulse currents

i_p = polarization currents

I_k = conductive currents

$$\tan \delta = \frac{\text{Active Power}}{\text{Re active Power}} = \frac{u.i_{ins}.\cos \phi}{u.i_{ins}.\sin \phi} = \frac{{}^i R(\text{Total})}{{}^i C(\text{Total})}$$

The active part of total insulation current through the capacitor is $\omega C U \tan \delta$. Hence the active power loss $'P_{ac}'$ is given by

$$P_{ac} = \omega.C.U^2. \tan \delta$$

Where C is total effective capacitance of the test object: $\tan \delta$ is a function of temperature, frequency and magnitude of the applied voltage U.

Measurement of Partial Breakdown (PB) in Dielectrics

Fig. xvi A PB free 100 kV Transformer and 1.1 nF Coupling Capacitor
setup in HV Laboratory at IIT Kanpur

PB in solid dielectrics are most damaging. These may reduce the life of the equipment drastically. The damage not only depends upon the intensity of PB but also upon the location and the time required for the formation of conducting path (treeing process). A Partial Breakdown Detector is required along with a PB free experimental set up at the required voltage. The measurement of quantities e.g. PB inception, PB extinction voltages and the level of permissible discharge at a given voltage are specified in BIS6209 and IEC-270. Besides, location of PB in any electrical equipment e.g. cables, transformers, electrical machines or GIS requires special techniques.

There are three basic circuits which can be employed for the measurement of PB measurable quantities suitable to convenience or specific requirement. These circuit are shown in Fig. xvii (a), (b), (c). As it can seen this figure, the coupling capacitor C_c , Measuring impedence Z_m are essential part of the circuit in all the three methods. C_1 and C_2 comprise of capacitive voltage divider for the measurement of the test voltage applied on the test object C_{Test}.

If a global breakdown or complete rupture of dielectric is expected to take place during the PB measurements, it is advisable to chose the circuit shown in Fig. xvii (a) inorder to protect the measuring impedence Z_m and the voltage measuring unit from over-voltages developed at breakdown.

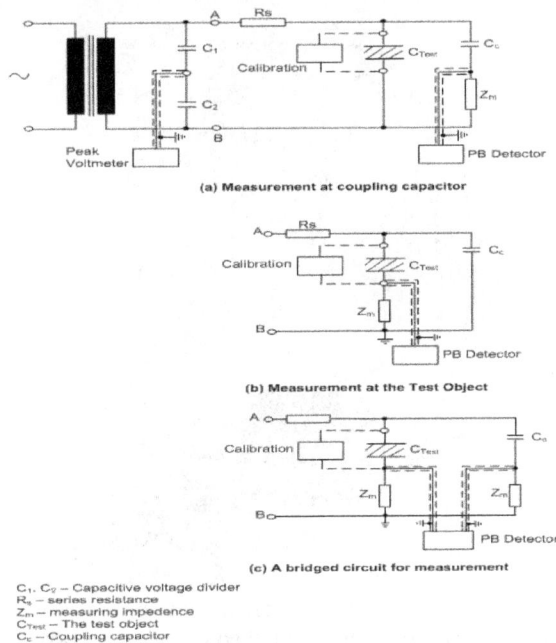

(a) Measurement at coupling capacitor

(b) Measurement at the Test Object

(c) A bridged circuit for measurement

C_1, C_2 – Capacitive voltage divider
R_s – series resistance
Z_m – measuring impedence
C_{Test} – The test object
C_c – Coupling capacitor

Fig. xvii Partial Breakdown Measurement Circuits

If the capacitance of the coupling capacitor C_c is smaller than the capacitance of the test object, the sensitivity of the measurement with circuit in Fig. xvii (b) is higher than with the circuit in (a).

The bridge circuit, shown in Fig. xvii (c), is more suitable for the measurement at sites having high electromagnetic interference (EMI). Some times it is difficult to make distinction between the PB pulse generated within the test object and the external interference. However, incorporation of suitable filters in modern PB detection facilitate the measurement even under external EMI. Calibration of the measuring circuit is required to be done with "standard calibrating charge generator". The measurable quantity of the apparent charge and the sensitivity of the measurement depend largely upon the calibration.

Transformers: A Comprehensive Study

A transformer conducts electricity from one place to another through electromagnetic induction. It allows electrical current to be adjusted and regulated. The aspects elucidated in this chapter are of vital importance, and provide a better understanding of high voltage engineering.

Transformer

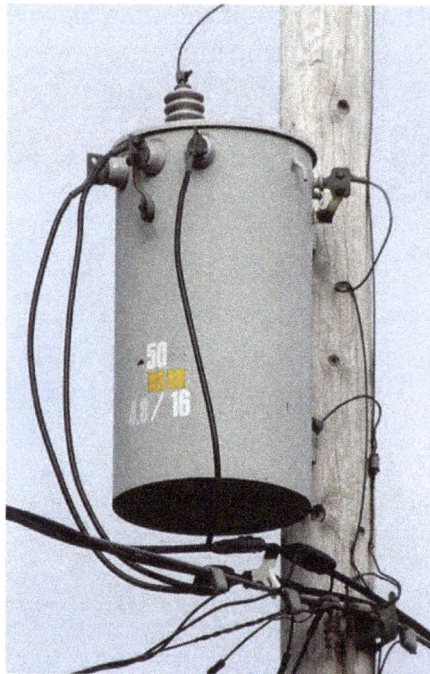

Pole-mounted distribution transformer with center-tapped secondary winding used to provide "split-phase" power for residential and light commercial service, which in North America is typically rated 120/240 V.

A transformer is an electrical device that transfers electrical energy between two or more circuits through electromagnetic induction. A varying current in one coil of the transformer produces a varying magnetic field, which in turn induces a voltage in a second coil. Power can be transferred between the two coils through the magnetic field, without a metallic connection between the two circuits. Faraday's law of induction discovered in 1831 described this effect. Transformers are used to increase or decrease the alternating voltages in electric power applications.

Since the invention of the first constant-potential transformer in 1885, transformers have become essential for the transmission, distribution, and utilization of alternating current electrical energy. A wide range of transformer designs is encountered in electronic and electric power applications.

Transformers range in size from RF transformers less than a cubic centimeter in volume to units interconnecting the power grid weighing hundreds of tons.

Basic Principles

Ideal Transformer Equations (eq.)

By Faraday's law of induction:

$$V_S = -N_S \frac{d\Phi}{dt} \tag{1}$$

$$V_P = -N_P \frac{d\Phi}{dt} \tag{2}$$

Combining ratio of (1) & (2)

Turns ratio $= \dfrac{V_P}{V_S} = \dfrac{-N_P}{-N_S} = a \; ... \; (3)$ where

      for step-down transformers, $a > 1$

      for step-up transformers, $a < 1$

By law of conservation of energy, apparent, real and reactive power are each conserved in the input and output

$$S = I_P V_P = I_S V_S \tag{4}$$

Combining (3) & (4) with this endnote yields the ideal transformer identity

$$\frac{V_P}{V_S} = \frac{I_S}{I_P} = \frac{N_P}{N_S} = \sqrt{\frac{L_P}{L_S}} = a. \tag{5}$$

By Ohm's law and ideal transformer identity

$$Z_L = \frac{V_S}{I_S} \tag{6}$$

Apparent load impedance Z'_L (Z_L referred to the primary)

$$Z'_L = \frac{V_P}{I_P} = \frac{aV_S}{I_S / a} = a^2 \frac{V_S}{I_S} = a^2 Z_L \tag{7}$$

For simplification or approximation purposes, it is very common to analyze the transformer as an ideal transformer model as presented in the two images. An ideal transformer is a theoretical, linear transformer that is lossless and perfectly coupled; that is, there are no energy losses and flux is completely confined within the magnetic core. Perfect coupling implies infinitely high core magnetic permeability and winding inductances and zero net magnetomotive force.

$$a{:}1$$
$$N_P{:}N_S$$

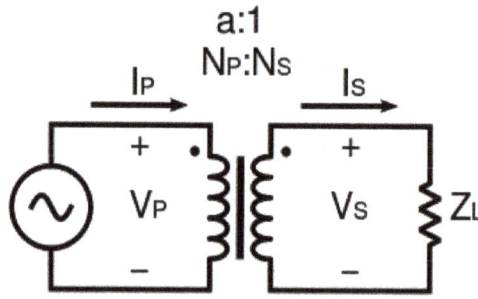

Ideal transformer connected with source V_p on primary and load impedance
Z_L on secondary, where $0 < Z_L < \infty$.

A varying current in the transformer's primary winding creates a varying magnetic flux in the transformer core and a varying magnetic field impinging on the secondary winding. This varying magnetic field at the secondary winding induces a varying EMF or voltage in the secondary winding due to electromagnetic induction. The primary and secondary windings are wrapped around a core of infinitely high magnetic permeability so that all of the magnetic flux passes through both the primary and secondary windings. With a voltage source connected to the primary winding and load impedance connected to the secondary winding, the transformer currents flow in the indicated directions.

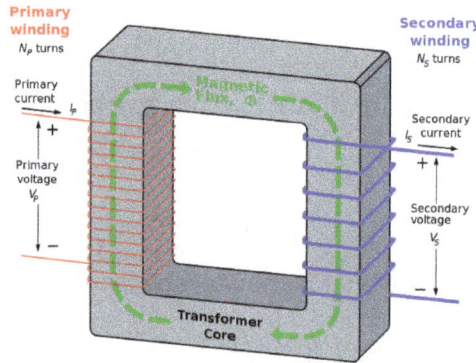

Ideal transformer and induction law

According to Faraday's law, since the same magnetic flux passes through both the primary and secondary windings in an ideal transformer, a voltage is induced in each winding, according to eq. (1) in the secondary winding case, according to eq. in the primary winding case. The primary EMF is sometimes termed counter EMF. This is in accordance with Lenz's law, which states that induction of EMF always opposes development of any such change in magnetic field.

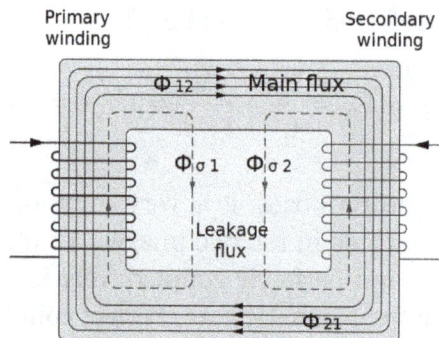

Leakage flux of a transformer

The transformer winding voltage ratio is thus shown to be directly proportional to the winding turns ratio according to eq. (3). common usage having evolved over time from 'turn ratio' to 'turns ratio'. However, some sources use the inverse definition.

According to the law of conservation of energy, any load impedance connected to the ideal transformer's secondary winding results in conservation of apparent, real and reactive power consistent with eq. (4).

The ideal transformer identity shown in eq. (5) is a reasonable approximation for the typical commercial transformer, with voltage ratio and winding turns ratio both being inversely proportional to the corresponding current ratio.

By Ohm's law and the ideal transformer identity:

- the secondary circuit load impedance can be expressed as eq. (6).
- the apparent load impedance *referred* to the primary circuit is derived in eq. (7) to be equal to the turns ratio squared times the secondary circuit load impedance.

Real Transformer

Deviations from Ideal

The ideal transformer model neglects the following basic linear aspects in real transformers:

a) Core losses, collectively called magnetizing current losses, consisting of

- Hysteresis losses due to nonlinear application of the voltage applied in the transformer core, and
- Eddy current losses due to joule heating in the core that are proportional to the square of the transformer's applied voltage.

b) Whereas windings in the ideal model have no resistances and infinite inductances, the windings in a real transformer have finite non-zero resistances and inductances associated with:

- Joule losses due to resistance in the primary and secondary windings.
- Leakage flux that escapes from the core and passes through one winding only resulting in primary and secondary reactive impedance.

Leakage Flux

The ideal transformer model assumes that all flux generated by the primary winding links all the turns of every winding, including itself. In practice, some flux traverses paths that take it outside the windings. Such flux is termed *leakage flux*, and results in leakage inductance in series with the mutually coupled transformer windings. Leakage flux results in energy being alternately stored in and discharged from the magnetic fields with each cycle of the power supply. It is not directly a power loss, but results in inferior voltage regulation, causing the secondary voltage not to be directly proportional to the primary voltage, particularly under heavy load. Transformers are therefore normally designed to have very low leakage inductance.

In some applications increased leakage is desired, and long magnetic paths, air gaps, or magnetic bypass shunts may deliberately be introduced in a transformer design to limit the short-circuit current it will supply. Leaky transformers may be used to supply loads that exhibit negative resistance, such as electric arcs, mercury- and sodium- vapor lamps and neon signs or for safely handling loads that become periodically short-circuited such as electric arc welders.

Air gaps are also used to keep a transformer from saturating, especially audio-frequency transformers in circuits that have a DC component flowing in the windings.

Knowledge of leakage inductance is also useful when transformers are operated in parallel. It can be shown that if the percent impedance and associated winding leakage reactance-to-resistance (X/R) ratio of two transformers were hypothetically exactly the same, the transformers would share power in proportion to their respective volt-ampere ratings (e.g. 500 kVA unit in parallel with 1,000 kVA unit, the larger unit would carry twice the current). However, the impedance tolerances of commercial transformers are significant. Also, the Z impedance and X/R ratio of different capacity transformers tends to vary, corresponding 1,000 kVA and 500 kVA units' values being, to illustrate, respectively, $Z \approx 5.75\%$, $X/R \approx 3.75$ and $Z \approx 5\%$, $X/R \approx 4.75$.

Equivalent Circuit

Referring to the diagram, a practical transformer's physical behavior may be represented by an equivalent circuit model, which can incorporate an ideal transformer.

Winding joule losses and leakage reactances are represented by the following series loop impedances of the model:

- Primary winding: R_P, X_P

- Secondary winding: R_S, X_S.

In normal course of circuit equivalence transformation, R_S and X_S are in practice usually referred to the primary side by multiplying these impedances by the turns ratio squared, $(N_P/N_S)^2 = a^2$.

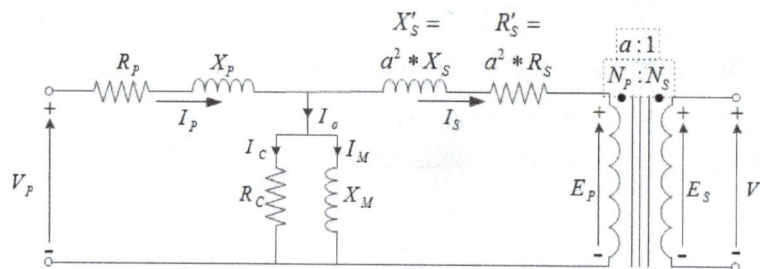

Real transformer equivalent circuit

Core loss and reactance is represented by the following shunt leg impedances of the model:

- Core or iron losses: R_C

- Magnetizing reactance: X_M.

R_C and X_M are collectively termed the *magnetizing branch* of the model.

Core losses are caused mostly by hysteresis and eddy current effects in the core and are proportional to the square of the core flux for operation at a given frequency. The finite permeability core requires a magnetizing current I_M to maintain mutual flux in the core. Magnetizing current is in phase with the flux, the relationship between the two being non-linear due to saturation effects. However, all impedances of the equivalent circuit shown are by definition linear and such non-linearity effects are not typically reflected in transformer equivalent circuits. With sinusoidal supply, core flux lags the induced EMF by 90°. With open-circuited secondary winding, magnetizing branch current I_0 equals transformer no-load current.

Instrument transformer, with polarity dot and X1 markings on LV side terminal

The resulting model, though sometimes termed 'exact' equivalent circuit based on linearity assumptions, retains a number of approximations. Analysis may be simplified by assuming that magnetizing branch impedance is relatively high and relocating the branch to the left of the primary impedances. This introduces error but allows combination of primary and referred secondary resistances and reactances by simple summation as two series impedances.

Transformer equivalent circuit impedance and transformer ratio parameters can be derived from the following tests: open-circuit test, short-circuit test, winding resistance test, and transformer ratio test.

Basic Transformer Parameters and Construction

Polarity

A dot convention is often used in transformer circuit diagrams, nameplates or terminal markings to define the relative polarity of transformer windings. Positively increasing instantaneous current entering the primary winding's dot end induces positive polarity voltage at the secondary winding's dot end.

Effect of Frequency

By Faraday's law of induction shown in eq. (1) and (2), transformer EMFs vary according to the derivative of flux with respect to time. The ideal transformer's core behaves linearly with time for any non-zero frequency. Flux in a real transformer's core behaves non-linearly in relation to magnetization current as the instantaneous flux increases beyond a finite linear range resulting in magnetic saturation associated with increasingly large magnetizing current, which eventually leads to transformer overheating.

Transformer Universal EMF Equation

If the flux in the core is purely sinusoidal, the relationship for either winding between its rms voltage E_{rms} of the winding, and the supply frequency f, number of turns N, core cross-sectional area a in m² and peak magnetic flux density B_{peak} in Wb/m² or T (tesla) is given by the universal EMF equation:

$$E_{rms} = \frac{2\pi f N a B_{peak}}{\sqrt{2}} \approx 4.44 f N a B_{peak}$$

If the flux does not contain even harmonics the following equation can be used for half-cycle average voltage E_{avg} of any waveshape:

$$E_{avg} = 4 f N a B_{peak}$$

The EMF of a transformer at a given flux density increases with frequency. By operating at higher frequencies, transformers can be physically more compact because a given core is able to transfer more power without reaching saturation and fewer turns are needed to achieve the same impedance. However, properties such as core loss and conductor skin effect also increase with frequency. Aircraft and military equipment employ 400 Hz power supplies which reduce core and winding weight. Conversely, frequencies used for some railway electrification systems were much lower (e.g. 16.7 Hz and 25 Hz) than normal utility frequencies (50–60 Hz) for historical reasons concerned mainly with the limitations of early electric traction motors. Consequently, the transformers used to step-down the high overhead line voltages (e.g. 15 kV) were much larger and heavier for the same power rating than those required for the higher frequencies.

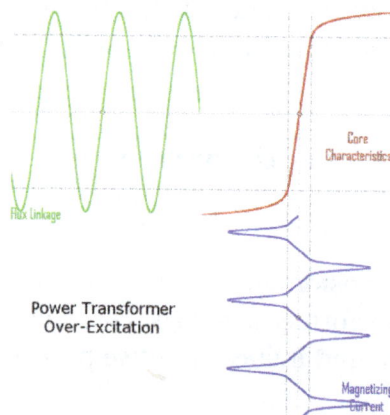

Power transformer over-excitation condition caused by decreased frequency; flux (green), iron core's magnetic characteristics (red) and magnetizing current (blue).

Operation of a transformer at its designed voltage but at a higher frequency than intended will lead to reduced magnetizing current. At a lower frequency, the magnetizing current will increase. Operation of a transformer at other than its design frequency may require assessment of voltages, losses, and cooling to establish if safe operation is practical. For example, transformers may need to be equipped with 'volts per hertz' over-excitation, ANSI function 24, relays to protect the transformer from overvoltage at higher than rated frequency.

One example is in traction transformers used for electric multiple unit and high-speed train service operating across regions with different electrical standards. The converter equipment and traction transformers have to accommodate different input frequencies and voltage (ranging from as high as 50 Hz down to 16.7 Hz and rated up to 25 kV) while being suitable for multiple AC asynchronous motor and DC converters and motors with varying harmonics mitigation filtering requirements.

Large power transformers are vulnerable to insulation failure due to transient voltages with high-frequency components, such as caused in switching or by lightning.

At much higher frequencies the transformer core size required drops dramatically: a physically small and cheap transformer can handle power levels that would require a massive iron core at mains frequency. The development of switching power semiconductor devices and complex integrated circuits made switch-mode power supplies viable, to generate a high frequency from a much lower one (or DC), change the voltage level with a small transformer, and, if necessary, rectify the changed voltage.

Energy Losses

Real transformer energy losses are dominated by winding resistance joule and core losses. Transformers' efficiency tends to improve with increasing transformer capacity. The efficiency of typical distribution transformers is between about 98 and 99 percent.

As transformer losses vary with load, it is often useful to express these losses in terms of no-load loss, full-load loss, half-load loss, and so on. Hysteresis and eddy current losses are constant at all load levels and dominate overwhelmingly without load, while variable winding joule losses dominating increasingly as load increases. The no-load loss can be significant, so that even an idle transformer constitutes a drain on the electrical supply. Designing energy efficient transformers for lower loss requires a larger core, good-quality silicon steel, or even amorphous steel for the core and thicker wire, increasing initial cost. The choice of construction represents a trade-off between initial cost and operating cost.

Transformer losses arise from:

Winding Joule Losses

Current flowing through a winding's conductor causes joule heating. As frequency increases, skin effect and proximity effect causes the winding's resistance and, hence, losses to increase.

Core Losses

Hysteresis Losses

Each time the magnetic field is reversed, a small amount of energy is lost due to hysteresis within the core. According to Steinmetz's formula, the heat energy due to hysteresis is given by

$W_h \approx \eta \beta_{max}^{1.6}$, and,

hysteresis loss is thus given by

$P_h \approx W_h f \approx \eta f \beta_{max}^{1.6}$

where, f is the frequency, η is the hysteresis coefficient and β_{max} is the maximum flux density, the empirical exponent of which varies from about 1.4 to 1.8 but is often given as 1.6 for iron.

Eddy Current Losses

Ferromagnetic materials are also good conductors and a core made from such a material also constitutes a single short-circuited turn throughout its entire length. Eddy currents therefore circulate within the core in a plane normal to the flux, and are responsible for resistive heating of the core material. The eddy current loss is a complex function of the square of supply frequency and inverse square of the material thickness. Eddy current losses can be reduced by making the core of a stack of plates electrically insulated from each other, rather than a solid block; all transformers operating at low frequencies use laminated or similar cores.

Magnetostriction Related Transformer Hum

Magnetic flux in a ferromagnetic material, such as the core, causes it to physically expand and contract slightly with each cycle of the magnetic field, an effect known as magnetostriction, the frictional energy of which produces an audible noise known as mains hum or transformer hum. This transformer hum is especially objectionable in transformers supplied at power frequencies and in high-frequency flyback transformers associated with television CRTs.

Stray Losses

Leakage inductance is by itself largely lossless, since energy supplied to its magnetic fields is returned to the supply with the next half-cycle. However, any leakage flux that intercepts nearby conductive materials such as the transformer's support structure will give rise to eddy currents and be converted to heat. There are also radiative losses due to the oscillating magnetic field but these are usually small.

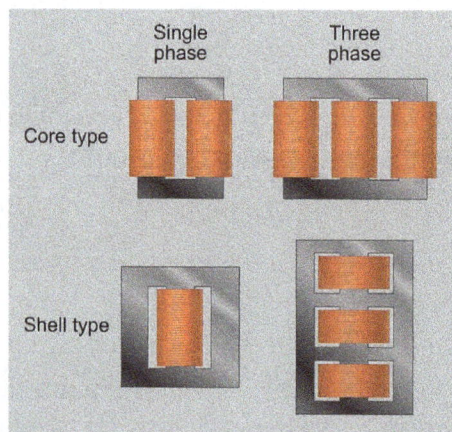

Core form = core type; shell form = shell type

Mechanical Vibration and Audible Noise Transmission

In addition to magnetostriction, the alternating magnetic field causes fluctuating forces between the primary and secondary windings. This energy incites vibration transmission in interconnected metalwork, thus amplifying audible transformer hum.

Core Form and Shell Form Transformers

Closed-core transformers are constructed in 'core form' or 'shell form'. When windings surround the core, the transformer is core form; when windings are surrounded by the core, the transformer is shell form. Shell form design may be more prevalent than core form design for distribution transformer applications due to the relative ease in stacking the core around winding coils. Core form design tends to, as a general rule, be more economical, and therefore more prevalent, than shell form design for high voltage power transformer applications at the lower end of their voltage and power rating ranges (less than or equal to, nominally, 230 kV or 75 MVA). At higher voltage and power ratings, shell form transformers tend to be more prevalent. Shell form design tends to be preferred for extra-high voltage and higher MVA applications because, though more labor-intensive to manufacture, shell form transformers are characterized as having inherently better kVA-to-weight ratio, better short-circuit strength characteristics and higher immunity to transit damage.

Construction

Laminated Steel Cores

Laminated core transformer showing edge of laminations at top of photo

Transformers for use at power or audio frequencies typically have cores made of high permeability silicon steel. The steel has a permeability many times that of free space and the core thus serves to greatly reduce the magnetizing current and confine the flux to a path which closely couples the windings. Early transformer developers soon realized that cores constructed from solid iron resulted in prohibitive eddy current losses, and their designs mitigated this effect with cores consisting of bundles of insulated iron wires. Later designs constructed the core by stacking layers of thin steel laminations, a principle that has remained in use. Each lamination is insulated from its neighbors by a thin non-conducting layer of insulation. The transformer universal EMF equation implies an acceptably large core cross-sectional area in order to avoid saturation.

The effect of laminations is to confine eddy currents to highly elliptical paths that enclose little flux, and so reduce their magnitude. Thinner laminations reduce losses, but are more laborious and expensive to construct. Thin laminations are generally used on high-frequency transformers, with some of very thin steel laminations able to operate up to 10 kHz.

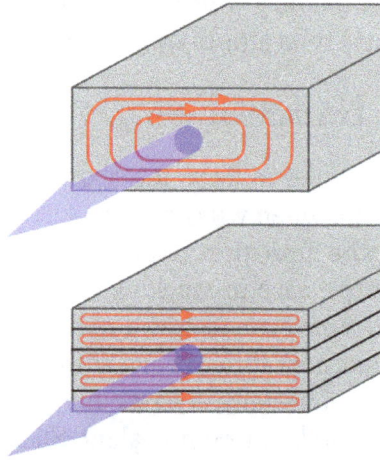

Laminating the core greatly reduces eddy-current losses

One common design of laminated core is made from interleaved stacks of E-shaped steel sheets capped with I-shaped pieces, leading to its name of 'E-I transformer'. Such a design tends to exhibit more losses, but is very economical to manufacture. The cut-core or C-core type is made by winding a steel strip around a rectangular form and then bonding the layers together. It is then cut in two, forming two C shapes, and the core assembled by binding the two C halves together with a steel strap. They have the advantage that the flux is always oriented parallel to the metal grains, reducing reluctance.

A steel core's remanence means that it retains a static magnetic field when power is removed. When power is then reapplied, the residual field will cause a high inrush current until the effect of the remaining magnetism is reduced, usually after a few cycles of the applied AC waveform. Overcurrent protection devices such as fuses must be selected to allow this harmless inrush to pass. On transformers connected to long, overhead power transmission lines, induced currents due to geomagnetic disturbances during solar storms can cause saturation of the core and operation of transformer protection devices.

Distribution transformers can achieve low no-load losses by using cores made with low-loss high-permeability silicon steel or amorphous (non-crystalline) metal alloy. The higher initial cost of the core material is offset over the life of the transformer by its lower losses at light load.

Solid Cores

Powdered iron cores are used in circuits such as switch-mode power supplies that operate above mains frequencies and up to a few tens of kilohertz. These materials combine high magnetic permeability with high bulk electrical resistivity. For frequencies extending beyond the VHF band, cores made from non-conductive magnetic ceramic materials called ferrites are common. Some radio-frequency transformers also have movable cores (sometimes called 'slugs') which allow adjustment of the coupling coefficient (and bandwidth) of tuned radio-frequency circuits.

Toroidal Cores

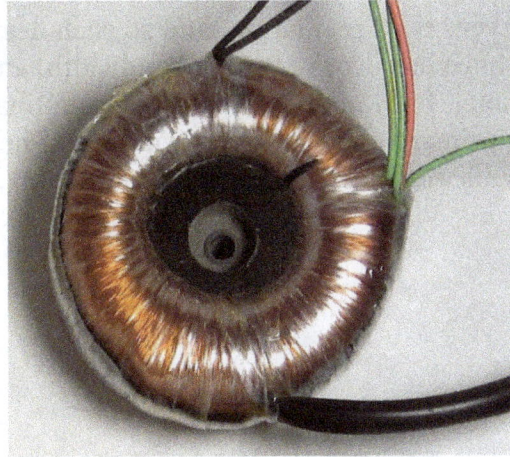

Small toroidal core transformer

Toroidal transformers are built around a ring-shaped core, which, depending on operating frequency, is made from a long strip of silicon steel or permalloy wound into a coil, powdered iron, or ferrite. A strip construction ensures that the grain boundaries are optimally aligned, improving the transformer's efficiency by reducing the core's reluctance. The closed ring shape eliminates air gaps inherent in the construction of an E-I core. The cross-section of the ring is usually square or rectangular, but more expensive cores with circular cross-sections are also available. The primary and secondary coils are often wound concentrically to cover the entire surface of the core. This minimizes the length of wire needed and provides screening to minimize the core's magnetic field from generating electromagnetic interference.

Toroidal transformers are more efficient than the cheaper laminated E-I types for a similar power level. Other advantages compared to E-I types, include smaller size (about half), lower weight (about half), less mechanical hum (making them superior in audio amplifiers), lower exterior magnetic field (about one tenth), low off-load losses (making them more efficient in standby circuits), single-bolt mounting, and greater choice of shapes. The main disadvantages are higher cost and limited power capacity. Because of the lack of a residual gap in the magnetic path, toroidal transformers also tend to exhibit higher inrush current, compared to laminated E-I types.

Ferrite toroidal cores are used at higher frequencies, typically between a few tens of kilohertz to hundreds of megahertz, to reduce losses, physical size, and weight of inductive components. A drawback of toroidal transformer construction is the higher labor cost of winding. This is because it is necessary to pass the entire length of a coil winding through the core aperture each time a single turn is added to the coil. As a consequence, toroidal transformers rated more than a few kVA are uncommon. Relatively few toroids are offered with power ratings above 10 kVA, and practically none above 25 kVA. Small distribution transformers may achieve some of the benefits of a toroidal core by splitting it and forcing it open, then inserting a bobbin containing primary and secondary windings.

Air Cores

A physical core is not an absolute requisite and a functioning transformer can be produced simply by placing the windings near each other, an arrangement termed an "air-core" transformer. The air which comprises the magnetic circuit is essentially lossless, and so an air-core transformer eliminates loss due to hysteresis in the core material. The leakage inductance is inevitably high, resulting in very poor regulation, and so such designs are unsuitable for use in power distribution. They have however very high bandwidth, and are frequently employed in radio-frequency applications, for which a satisfactory coupling coefficient is maintained by carefully overlapping the primary and secondary windings. They're also used for resonant transformers such as Tesla coils where they can achieve reasonably low loss in spite of the high leakage inductance.

Windings

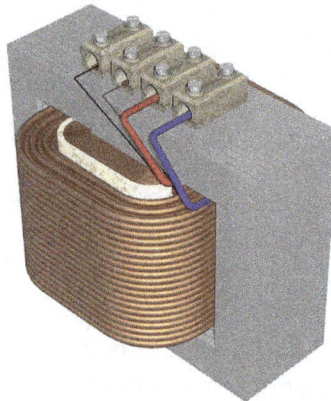

Windings are usually arranged concentrically to minimize flux leakage

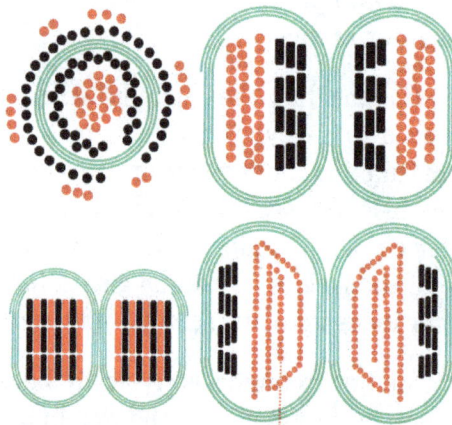

Cut view through transformer windings

High-frequency transformers operating in the tens to hundreds of kilohertz often have windings made of braided Litz wire to minimize the skin-effect and proximity effect losses. Large power transformers use multiple-stranded conductors as well, since even at low power frequencies non-uniform distribution of current would otherwise exist in high-current windings. Each strand is individually insulated, and the strands are arranged so that at certain points in the winding, or throughout the whole winding, each portion occupies different relative positions in the complete conductor. The transposition equalizes the current flowing in each strand of the conductor, and reduces eddy current losses in the winding itself. The stranded conductor is also more flexible than a solid conductor of similar size, aiding manufacture.

The windings of signal transformers minimize leakage inductance and stray capacitance to improve high-frequency response. Coils are split into sections, and those sections interleaved between the sections of the other winding.

Power-frequency transformers may have taps at intermediate points on the winding, usually on the higher voltage winding side, for voltage adjustment. Taps may be manually reconnected, or a manual or automatic switch may be provided for changing taps. Automatic on-load tap changers are used in electric power transmission or distribution, on equipment such as arc furnace transformers, or for automatic voltage regulators for sensitive loads. Audio-frequency transformers, used for the distribution of audio to public address loudspeakers, have taps to allow adjustment of impedance to each speaker. A center-tapped transformer is often used in the output stage of an audio power amplifier in a push-pull circuit. Modulation transformers in AM transmitters are very similar.

Dry-type transformer winding insulation systems can be either of standard open-wound 'dip-and-bake' construction or of higher quality designs that include vacuum pressure impregnation (VPI), vacuum pressure encapsulation (VPE), and cast coil encapsulation processes. In the VPI process, a combination of heat, vacuum and pressure is used to thoroughly seal, bind, and eliminate entrained air voids in the winding polyester resin insulation coat layer, thus increasing resistance to corona. VPE windings are similar to VPI windings but provide more protection against environmental effects, such as from water, dirt or corrosive ambients, by multiple dips including typically in terms of final epoxy coat.

Regarding image at top captioned, Cut view of transformer windings:

> The conducting material used for the windings depends upon the application, but in all cases the individual turns must be electrically insulated from each other to ensure that the current travels throughout every turn. For small power and signal transformers, in which currents are low and the potential difference between adjacent turns is small, the coils are often wound from enamelled magnet wire, such as Formvar wire. Larger power transformers operating at high voltages may be wound with copper rectangular strip conductors insulated by oil-impregnated paper and blocks of pressboard.

Legend

> White: Air, liquid or other insulating medium in conjunction with varnish, paper or other coil insulation.

> Green spiral: Grain oriented silicon steel.

> Black: Primary winding (Aluminum or copper).

> Red: Secondary winding (Aluminum or copper).

Cooling

To place the cooling problem in perspective, the accepted rule of thumb is that the life expectancy of insulation in all electrics, including all transformers, is halved for about every 7 °C to 10 °C increase in operating temperature, this life expectancy halving rule holding more narrowly when the increase is between about 7 °C to 8 °C in the case of transformer winding cellulose insulation.

Cutaway view of liquid-immersed construction transformer. The conservator (reservoir) at top provides liquid-to-atmosphere isolation as coolant level and temperature changes. The walls and fins provide required heat dissipation balance.

Small dry-type and liquid-immersed transformers are often self-cooled by natural convection and radiation heat dissipation. As power ratings increase, transformers are often cooled by forced-air cooling, forced-oil cooling, water-cooling, or combinations of these. Large transformers are filled with transformer oil that both cools and insulates the windings. Transformer oil is a highly refined mineral oil that cools the windings and insulation by circulating within the transformer tank. The mineral oil and paper insulation system has been extensively studied and used for more than 100 years. It is estimated that 50% of power transformers will survive 50 years of use, that the average age of failure of power transformers is about 10 to 15 years, and that about 30% of power transformer failures are due to insulation and overloading failures. Prolonged operation at elevated temperature degrades insulating properties of winding insulation and dielectric coolant, which not only shortens transformer life but can ultimately lead to catastrophic transformer failure. With a great body of empirical study as a guide, transformer oil testing including dissolved gas analysis provides valuable maintenance information. This underlines the need to monitor, model, forecast and manage oil and winding conductor insulation temperature conditions under varying, possibly difficult, power loading conditions.

Building regulations in many jurisdictions require indoor liquid-filled transformers to either use dielectric fluids that are less flammable than oil, or be installed in fire-resistant rooms. Air-cooled dry transformers can be more economical where they eliminate the cost of a fire-resistant transformer room.

The tank of liquid filled transformers often has radiators through which the liquid coolant circulates by natural convection or fins. Some large transformers employ electric fans for forced-air cooling, pumps for forced-liquid cooling, or have heat exchangers for water-cooling. An oil-immersed transformer may be equipped with a Buchholz relay, which, depending on severity of gas accumulation due to internal arcing, is used to either alarm or de-energize the transformer. Oil-immersed transformer installations usually include fire protection measures such as walls, oil containment, and fire-suppression sprinkler systems.

Polychlorinated biphenyls have properties that once favored their use as a dielectric coolant, though concerns over their environmental persistence led to a widespread ban on their use. Today, non-toxic, stable silicone-based oils, or fluorinated hydrocarbons may be used where the expense of a fire-resistant liquid offsets additional building cost for a transformer vault. PCBs for new equipment were banned in 1981 and in 2000 for use in existing equipment in United Kingdom Legislation enacted in Canada between 1977 and 1985 essentially bans PCB use in transformers manufactured in or imported into the country after 1980, the maximum allowable level of PCB contamination in existing mineral oil transformers being 50 ppm.

Some transformers, instead of being liquid-filled, have their windings enclosed in sealed, pressurized tanks and cooled by nitrogen or sulfur hexafluoride gas.

Experimental power transformers in the 500-to-1,000 kVA range have been built with liquid nitrogen or helium cooled superconducting windings, which eliminates winding losses without affecting core losses.

Insulation Drying

Construction of oil-filled transformers requires that the insulation covering the windings be thoroughly dried of residual moisture before the oil is introduced. Drying is carried out at the factory, and may also be required as a field service. Drying may be done by circulating hot air around the core, or by vapor-phase drying (VPD) where an evaporated solvent transfers heat by condensation on the coil and core.

For small transformers, resistance heating by injection of current into the windings is used. The heating can be controlled very well, and it is energy efficient. The method is called low-frequency heating (LFH) since the current used is at a much lower frequency than that of the power grid, which is normally 50 or 60 Hz. A lower frequency reduces the effect of inductance, so the voltage required can be reduced. The LFH drying method is also used for service of older transformers.

Bushings

Larger transformers are provided with high-voltage insulated bushings made of polymers or porcelain. A large bushing can be a complex structure since it must provide careful control of the electric field gradient without letting the transformer leak oil.

Classification Parameters

Transformers can be classified in many ways, such as the following:

- *Power capacity*: From a fraction of a volt-ampere (VA) to over a thousand MVA.

- *Duty of a transformer*: Continuous, short-time, intermittent, periodic, varying.

- *Frequency range*: Power-frequency, audio-frequency, or radio-frequency.

- *Voltage class*: From a few volts to hundreds of kilovolts.

- *Cooling type*: Dry and liquid-immersed – self-cooled, forced air-cooled; liquid-immersed – forced oil-cooled, water-cooled.

- *Circuit application*: Such as power supply, impedance matching, output voltage and current stabilizer or circuit isolation.

- *Utilization*: Pulse, power, distribution, rectifier, arc furnace, amplifier output, etc.

- *Basic magnetic form*: Core form, shell form, concentric, sandwich.

- *Constant-potential transformer descriptor*: Step-up, step-down, isolation.

- *General winding configuration*: By EIC vector group – various possible two-winding combinations of the phase designations delta, wye or star, and zigzag or interconnected star; other – autotransformer, Scott-T, zigzag grounding transformer winding.

- *Rectifier phase-shift winding configuration*: 2-winding, 6-pulse; 3-winding, 12-pulse; ... n-winding, [n-1]*6-pulse; polygon; etc..

Types

Various specific electrical application designs require a variety of transformer types. Although they all share the basic characteristic transformer principles, they are customize in construction or electrical properties for certain installation requirements or circuit conditions.

- *Autotransformer*: Transformer in which part of the winding is common to both primary and secondary circuits, leading to increased efficiency, smaller size, and a higher degree of voltage regulation.

- *Capacitor voltage transformer*: Transformer in which capacitor divider is used to reduce high voltage before application to the primary winding.

- *Distribution transformer, power transformer*: International standards make a distinction in terms of distribution transformers being used to distribute energy from transmission lines and networks for local consumption and power transformers being used to transfer electric energy between the generator and distribution primary circuits.

- *Phase angle regulating transformer*: A specialised transformer used to control the flow of real power on three-phase electricity transmission networks.

- *Scott-T transformer*: Transformer used for phase transformation from three-phase to two-phase and vice versa.

- *Polyphase transformer*: Any transformer with more than one phase.

- *Grounding transformer*: Transformer used for grounding three-phase circuits to create a neutral in a three wire system, using a wye-delta transformer, or more commonly, a zigzag grounding winding.

- *Leakage transformer*: Transformer that has loosely coupled windings.

- *Resonant transformer*: Transformer that uses resonance to generate a high secondary voltage.

- *Audio transformer*: Transformer used in audio equipment.

- *Output transformer*: Transformer used to match the output of a valve amplifier to its load.

- *Instrument transformer*: Potential or current transformer used to accurately and safely represent voltage, current or phase position of high voltage or high power circuits.

- *Pulse transformer*: Specialized small-signal transformer used to transmit digital signaling while providing electrical isolation, commonly used in Ethernet computer networks as 10BASE-T, 100BASE-T and 1000BASE-T.

An electrical substation in Melbourne, Australia showing three of five 220 kV – 66 kV transformers, each with a capacity of 150 MVA.

Applications

Transformers are used to increase (or step-up) voltage before transmitting electrical energy over long distances through wires. Wires have resistance which loses energy through joule heating at a rate corresponding to square of the current. By transforming power to a higher voltage transformers enable economical transmission of power and distribution. Consequently, transformers have shaped the electricity supply industry, permitting generation to be located remotely from points of demand. All but a tiny fraction of the world's electrical power has passed through a series of transformers by the time it reaches the consumer.

Transformer at the Limestone Generating Station in Manitoba, Canada

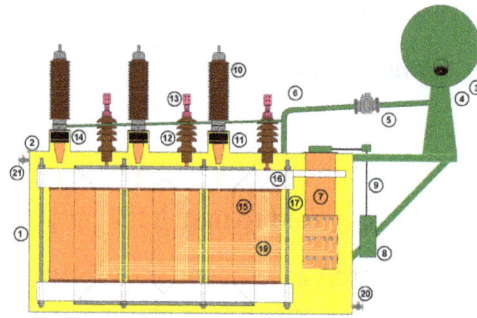

Schematic of a large oil filled power transformer

Since the high voltages carried in the wires are significantly greater than what is needed in-home, transformers are also used extensively in electronic products to decrease (or step-down) the supply voltage to a level suitable for the low voltage circuits they contain. The transformer also electrically isolates the end user from contact with the supply voltage.

Signal and audio transformers are used to couple stages of amplifiers and to match devices such as microphones and record players to the input of amplifiers. Audio transformers allowed telephone circuits to carry on a two-way conversation over a single pair of wires. A balun transformer converts a signal that is referenced to ground to a signal that has balanced voltages to ground, such as between external cables and internal circuits. Transformers made to medical grade standards isolate the users from the direct current. These are found commonly used in conjunction with hospital beds, dentist chairs, and other medical lab equipment.

History

Discovery of Induction

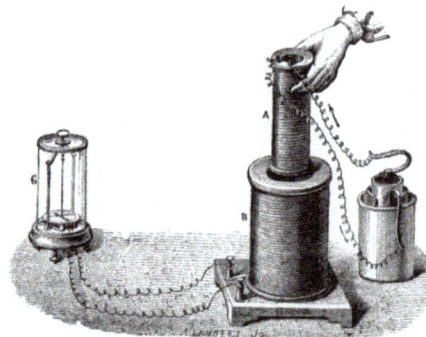

Faraday's experiment with induction between coils of wire

Electromagnetic induction, the principle of the operation of the transformer, was discovered independently by Michael Faraday in 1831, Joseph Henry in 1832, and others. The relationship between EMF and magnetic flux is an equation now known as Faraday's law of induction:

$$|\mathcal{E}| = \left|\frac{d\Phi_{B}}{dt}\right|.$$

where $|\mathcal{E}|$ is the magnitude of the EMF in Volts and Φ_{B} is the magnetic flux through the circuit in webers.

Faraday performed early experiments on induction between coils of wire, including winding a pair of coils around an iron ring, thus creating the first toroidal closed-core transformer. However he only applied individual pulses of current to his transformer, and never discovered the relation between the turns ratio and EMF in the windings.

Induction coil, 1900, Bremerhaven, Germany

Induction Coils

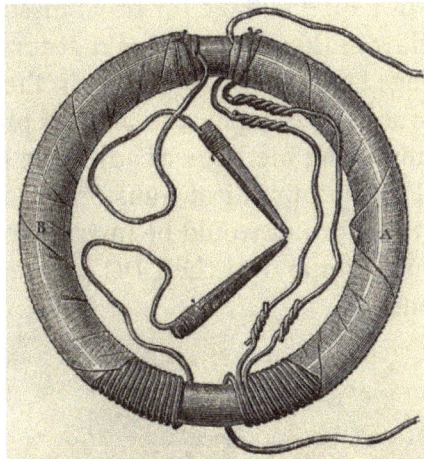

Faraday's ring transformer

The first type of transformer to see wide use was the induction coil, invented by Rev. Nicholas Callan of Maynooth College, Ireland in 1836. He was one of the first researchers to realize the more turns the secondary winding has in relation to the primary winding, the larger the induced secondary EMF will be. Induction coils evolved from scientists' and inventors' efforts to get higher voltages from batteries. Since batteries produce direct current (DC) rather than AC, induction coils relied upon vibrating electrical contacts that regularly interrupted the current in the primary to create the flux changes necessary for induction. Between the 1830s and the 1870s, efforts to build better induction coils, mostly by trial and error, slowly revealed the basic principles of transformers.

First Alternating Current Transformers

By the 1870s, efficient generators producing alternating current (AC) were available, and it was found AC could power an induction coil directly, without an interrupter.

In 1876, Russian engineer Pavel Yablochkov invented a lighting system based on a set of induction coils where the primary windings were connected to a source of AC. The secondary windings could be connected to several 'electric candles' (arc lamps) of his own design. The coils Yablochkov employed functioned essentially as transformers.

In 1878, the Ganz factory, Budapest, Hungary, began equipment for electric lighting and, by 1883, had installed over fifty systems in Austria-Hungary. Their AC systems used arc and incandescent lamps, generators, and other equipment.

Lucien Gaulard and John Dixon Gibbs first exhibited a device with an open iron core called a 'secondary generator' in London in 1882, then sold the idea to the Westinghouse company in the United States. They also exhibited the invention in Turin, Italy in 1884, where it was adopted for an electric lighting system.

Early Series Circuit Transformer Distribution

Induction coils with open magnetic circuits are inefficient at transferring power to loads. Until about 1880, the paradigm for AC power transmission from a high voltage supply to a low voltage load was a series circuit. Open-core transformers with a ratio near 1:1 were connected with their primaries in series to allow use of a high voltage for transmission while presenting a low voltage to the lamps. The inherent flaw in this method was that turning off a single lamp (or other electric device) affected the voltage supplied to all others on the same circuit. Many adjustable transformer designs were introduced to compensate for this problematic characteristic of the series circuit, including those employing methods of adjusting the core or bypassing the magnetic flux around part of a coil. Efficient, practical transformer designs did not appear until the 1880s, but within a decade, the transformer would be instrumental in the War of Currents, and in seeing AC distribution systems triumph over their DC counterparts, a position in which they have remained dominant ever since.

The ZBD team consisted of Károly Zipernowsky, Ottó Bláthy and Miksa Déri

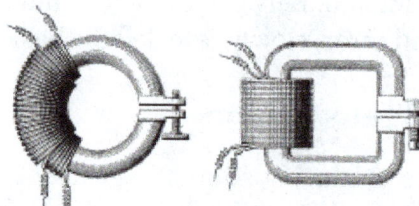

W. STANLEY, Jr.
INDUCTION COIL.

No. 349,611
Patented Sept. 21, 1886.

Stanley's 1886 design for adjustable gap open-core induction coils

Closed-core Transformers and Parallel Power Distribution

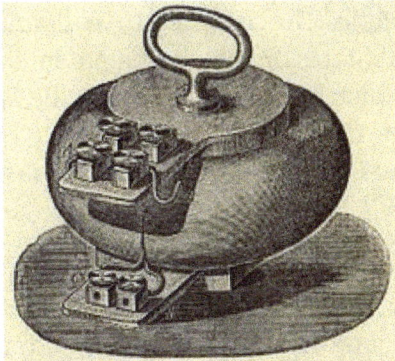

Shell form transformer. Sketch used by Uppenborn to describe ZBD engineers' 1885 patents and earliest articles.

Core form, front; shell form, back. Earliest specimens of ZBD-designed high-efficiency constant-potential transformers manufactured at the Ganz factory in 1885.

In the autumn of 1884, Károly Zipernowsky, Ottó Bláthy and Miksa Déri (ZBD), three engineers associated with the Ganz factory, had determined that open-core devices were impracticable, as they were incapable of reliably regulating voltage. In their joint 1885 patent applications for novel transformers (later called ZBD transformers), they described two designs with closed magnetic circuits where copper windings were either a) wound around iron wire ring core or b) surrounded by iron wire core. The two designs were the first application of the two basic transformer constructions in common use to this day, which can as a class all be termed as either core form or shell form (or alternatively, core type or shell type), as in a) or b), respectively. The Ganz factory had also in the autumn of 1884 made delivery of the world's first five high-efficiency AC transformers, the first of these units having been shipped on September 16, 1884. This first unit had been manufactured to the following specifications: 1,400 W, 40 Hz, 120:72 V, 11.6:19.4 A, ratio 1.67:1, one-phase, shell form.

In both designs, the magnetic flux linking the primary and secondary windings traveled almost entirely within the confines of the iron core, with no intentional path through air. The new transformers were 3.4 times more efficient than the open-core bipolar devices of Gaulard and Gibbs. The ZBD patents included two other major interrelated innovations: one concerning the use of parallel connected, instead of series connected, utilization loads, the other concerning the ability to have high turns ratio transformers such that the supply network voltage could be much higher (initially 1,400 to 2,000 V) than the voltage of utilization loads (100 V initially preferred). When employed in parallel connected electric distribution systems, closed-core transformers finally made it technically and economically feasible to provide electric power for lighting in homes, businesses and public spaces. Bláthy had suggested the use of closed cores, Zipernowsky had suggested the use of parallel shunt connections, and Déri had performed the experiments;

Transformers today are designed on the principles discovered by the three engineers. They also popularized the word 'transformer' to describe a device for altering the EMF of an electric current, although the term had already been in use by 1882. In 1886, the ZBD engineers designed, and the Ganz factory supplied electrical equipment for, the world's first power station that used AC generators to power a parallel connected common electrical network, the steam-powered Rome-Cerchi power plant.

Although George Westinghouse had bought Gaulard and Gibbs' patents in 1885, the Edison Electric Light Company held an option on the US rights for the ZBD transformers, requiring Westinghouse to pursue alternative designs on the same principles. He assigned to William Stanley the task of developing a device for commercial use in United States. Stanley's first patented design was for induction coils with single cores of soft iron and adjustable gaps to regulate the EMF present in the secondary winding. This design was first used commercially in the US in 1886 but Westinghouse was intent on improving the Stanley design to make it (unlike the ZBD type) easy and cheap to produce.

Westinghouse, Stanley and associates soon developed an easier to manufacture core, consisting of a stack of thin 'Eshaped' iron plates, insulated by thin sheets of paper or other insulating material. Prewound copper coils could then be slid into place, and straight iron plates laid in to create a closed magnetic circuit. Westinghouse applied for a patent for the new low-cost design in December 1886; it was granted in July 1887.

Other Early Transformer Designs

In 1889, Russian-born engineer Mikhail Dolivo-Dobrovolsky developed the first three-phase transformer at the Allgemeine Elektricitäts-Gesellschaft ('General Electricity Company') in Germany.

In 1891, Nikola Tesla invented the Tesla coil, an air-cored, dual-tuned resonant transformer for producing very high voltages at high frequency.

Audio frequency transformers ('repeating coils') were used by early experimenters in the development of the telephone.

Transformer Types

A variety of types of electrical transformer are made for different purposes. Despite their design differences, the various types employ the same basic principle as discovered in 1831 by Michael Faraday, and share several key functional parts.

An electric arc furnace the transformer has heavy copper bus for the low voltage winding, which can be rated for tens of thousands of amperes. They are immersed in oil for cooling and insulation, and are designed to survive frequent short circuits.

Power Transformers

Laminated Core

This is the most common type of transformer, widely used in electric power transmission and appliances to convert mains voltage to low voltage to power electronic devices. They are available in power ratings ranging from mW to MW. The insulated laminations minimizes eddy current losses in the iron core.

Small appliance and electronic transformers may use a split bobbin, giving a high level of insulation between the windings. The rectangular cores are made up of stampings, often in E-I shape pairs, but other shapes are sometimes used. Shields between primary and secondary may be fitted to reduce EMI (electromagnetic interference), or a screen winding is occasionally used.

Small appliance and electronics transformers may have a thermal cut-out built into the winding, to shut-off power at high temperatures to prevent further overheating.

Toroidal

Doughnut shaped toroidal transformers save space compared to E-I cores, and sometimes to reduce external magnetic field. These use a ring shaped core, copper windings wrapped round this ring (and thus threaded through the ring during winding), and tape for insulation.

Toroidal transformers have a lower external magnetic field compared to rectangular transformers, and can be smaller for a given power rating. However, they cost more to make, as winding requires more complex and slower equipment.

They can be mounted by a bolt through the center, using washers and rubber pads or by potting in resin.

Autotransformer

An autotransformer has one winding that is tapped at some point along the winding. Voltage is applied across a portion of the winding, and a higher (or lower) voltage is produced across another portion of the same winding. The equivalent power rating of the autotransfomer is lower than the actual load power rating. It is calculated by: load VA × (|Vin − Vout|)/Vin. For example, an auto transformer that adapts a 1000 VA load rated at 120 Volts to a 240 Volt supply has an equivalent rating of at least: 1,000VA × (240V − 120V) / 240V = 500VA. However, the actual rating (shown on the tally plate) must be at least 1000 VA.

For voltage ratios that don't exceed about 3:1, an autotransformer is cheaper, lighter, smaller, and more efficient than an isolating (two-winding) transformer of the same rating. Large three-phase autotransformers are used in electric power distribution systems, for example, to interconnect 33 kV and 66 kV sub-transmission networks.

Variable Autotransformer

By exposing part of the winding coils of an autotransformer, and making the secondary connection through a sliding carbon brush, an autotransformer with a near-continuously variable turns ratio can be obtained, allowing for wide voltage adjustment in very small increments.

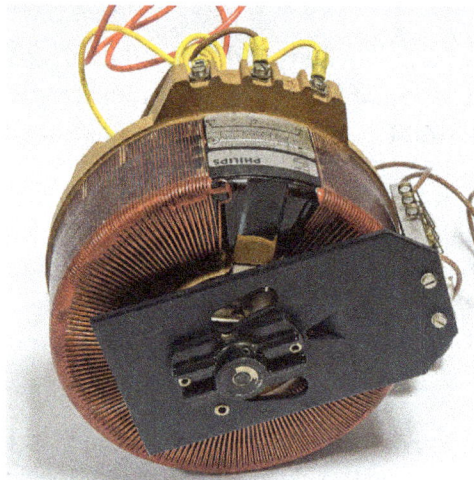

Variable autotransformer

Induction Regulator

The induction regulator is similar in design to a wound-rotor induction motor but it is essentially a transformer whose output voltage is varied by rotating its secondary relative to the primary—i.e., rotating the angular position of the rotor. It can be seen as a power transformer exploiting rotating magnetic fields. The major advantage of the induction regulator is that unlike variacs, they are practical for transformers over 5 kVA. Hence, such regulators find widespread use in high-voltage laboratories.

Polyphase Transformer

A high-voltage transformer being dismantled

For polyphase systems, multiple single-phase transformers can be used, or all phases can be connected to a single polyphase transformer. For a three phase transformer, the three primary windings are connected together and the three secondary windings are connected together. Examples of connections are wye-delta, delta-wye, delta-delta and wye-wye. A vector group indicates the configuration of the windings and the phase angle difference between them. If a winding is connected to earth (grounded), the earth connection point is usually the center point of a wye winding. If the secondary is a delta winding, the ground may be connected to a center tap on one winding (high leg delta) or one phase may be grounded (corner grounded delta). A special purpose polyphase transformer is the zigzag transformer. There are many possible configurations that may involve more or fewer than six windings and various tap connections.

Three-phase transformers 380kV/110kV and 110kV/20kV

Grounding Transformer

Grounding transformers let three wire (delta) polyphase system supplies accommodate phase to neutral loads by providing a return path for current to a neutral. Grounding transformers most commonly incorporate a single winding transformer with a zigzag winding configuration but may also be created with a wye-delta isolated winding transformer connection.

Phase-shifting Transformer

This is a specialized type of transformer which can be configured to adjust the phase relationship between input and output. This allows power flow in an electric grid to be controlled, e.g. to steer power flows away from a shorter (but overloaded) link to a longer path with excess capacity.

Variable-frequency Transformer

A *variable-frequency transformer* is a specialized three-phase power transformer which allows the phase relationship between the input and output windings to be continuously adjusted by rotating one half. They are used to interconnect electrical grids with the same nominal frequency but without synchronous phase coordination.

Leakage or Stray Field Transformers

A leakage transformer, also called a stray-field transformer, has a significantly higher leakage inductance than other transformers, sometimes increased by a magnetic bypass or shunt in its core between primary and secondary, which is sometimes adjustable with a set screw. This provides a transformer with an inherent current limitation due to the loose coupling between its primary and the secondary windings. In this case, it is short-circuit inductance which is actually acting as a current limiting parameter. The output and input currents are low enough to prevent thermal overload under all load conditions—even if the secondary is shorted.

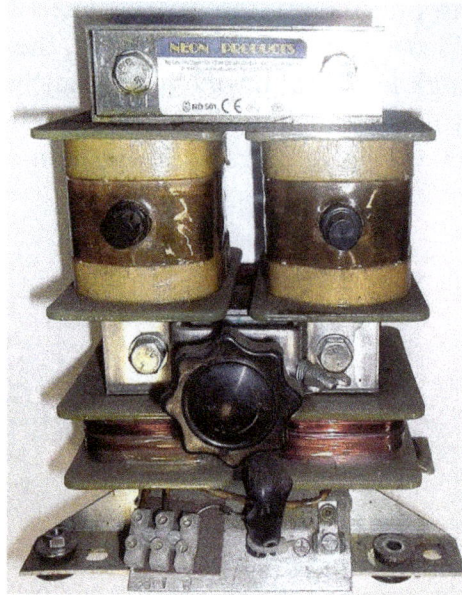

Leakage transformer

Uses

Leakage transformers are used for arc welding and high voltage discharge lamps (neon lights and cold cathode fluorescent lamps, which are series connected up to 7.5 kV AC). It acts then both as a voltage transformer and as a magnetic ballast.

Other applications are short-circuit-proof extra-low voltage transformers for toys or doorbell installations.

Resonant Transformer

A resonant transformer is a transformer in which one or both windings has a capacitor across it and functions as a tuned circuit. Used at radio frequencies, resonant transformers can function as high Q_factor bandpass filters. The transformer windings have either air or ferrite cores and the bandwidth can be adjusted by varying the coupling (mutual inductance). One common form is the IF (intermediate frequency) transformer, used in superheterodyne radio receivers. They are also used in radio transmitters.

Resonant transformers are also used in electronic ballasts for gas discharge lamps, and high voltage power supplies. They are also used in some types of switching power supplies. Here the short-circuit inductance value is an important parameter that determines the resonance frequency of the resonant transformer. Often only secondary winding has a resonant capacitor (or stray capacitance) and acts as a serial resonant tank circuit. When the short-circuit inductance of the secondary side of the transformer is L_{sc} and the resonant capacitor (or stray capacitance) of the secondary side is C_r, The resonance frequency ω_s of 1' is as follows:

$$\omega_s = \frac{1}{\sqrt{L_{sc}C_r}} = \frac{1}{\sqrt{(1-k^2)L_sC_r}}$$

The transformer is driven by a pulse or square wave for efficiency, generated by an electronic oscillator circuit. Each pulse serves to drive resonant sinusoidal oscillations in the tuned winding, and due to resonance a high voltage can be developed across the secondary.

Applications:

- Intermediate frequency (IF) transformer in superheterodyne radio receiver

- Tank transformers in radio transmitters

- Tesla coil

- CCFL inverter

- Oudin coil (or Oudin resonator; named after its inventor Paul Oudin)

- D'Arsonval apparatus

- Ignition coil or induction coil used in the ignition system of a petrol engine

- Electrical breakdown and insulation testing of high voltage equipment and cables. In the latter case, the transformer's secondary is resonated with the cable's capacitance.

Constant Voltage Transformer

By arranging particular magnetic properties of a transformer core, and installing a ferro-resonant tank circuit (a capacitor and an additional winding), a transformer can be arranged to automatically keep the secondary winding voltage relatively constant for varying primary supply without additional circuitry or manual adjustment. Ferro-resonant transformers run hotter than standard power transformers, because regulating action depends on core saturation, which reduces efficiency. The output waveform is heavily distorted unless careful measures are taken to prevent this. Saturating transformers provide a simple rugged method to stabilize an AC power supply.

Ferrite Core

Ferrite core power transformers are widely used in switched-mode power supplies (SMPSs). The powder core enables high-frequency operation, and hence much smaller size-to-power ratio than laminated-iron transformers.

Ferrite transformers are not used as power transformers at mains frequency since laminated iron cores cost less than an equivalent ferrite core.

Planar Transformer

Manufacturers etch spiral patterns on a printed circuit board to form the "windings" of a planar transformer, replacing the turns of wire used to make other types. Some planar transformers are commercially sold as discrete components. Other planar transformers are one of many components on a printed circuit board. A planar transformer can be thinner than other transformers, which is useful for low-profile applications or when several printed circuit boards are stacked. Almost all planar transformers use a ferrite planar core.

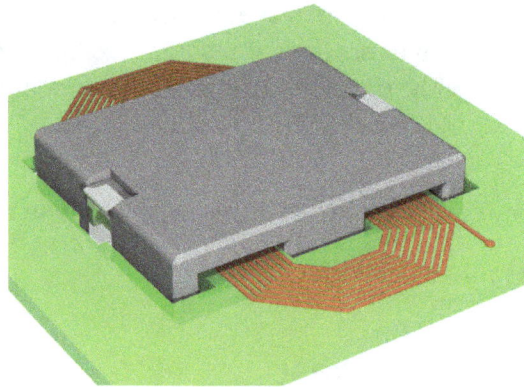

A planar transformer

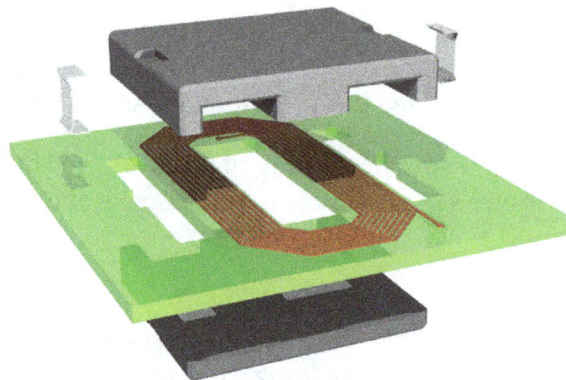

Exploded view: the spiral primary "winding" on one side of the PCB (the spiral secondary
"winding" is on the other side of the PCB).

Oil Cooled Transformer

Large transformers used in power distribution or electrical substations have their core and coils immersed in oil, which cools and insulates. Oil circulates through ducts in the coil and around the coil and core assembly, moved by convection. The oil is cooled by the outside of the tank in small ratings, and by an air-cooled radiator in larger ratings. Where a higher rating is required, or where the transformer is in a building or underground, oil pumps circulate the oil, and an oil-to-water heat exchanger may also be used. Some transformers may contain PCBs where or when its use was permitted. For example, until 1979 in South Africa. substitute fire-resistant liquids such as silicone oils are now used instead.

Cast Resin Transformer

Cast-resin power transformers encase the windings in epoxy resin. These transformers simplify installation since they are dry, without cooling oil, and so require no fire-proof vault for indoor installations. The epoxy protects the windings from dust and corrosive atmospheres. However, because the molds for casting the coils are only available in fixed sizes, the design of the transformers is less flexible, which may make them more costly if customized features (voltage, turns ratio, taps) are required.

Isolating Transformer

An isolation transformer links two circuits magnetically, but provides no metallic conductive path between the circuits. An example application would be in the power supply for medical equipment, when it is necessary to prevent any leakage from the AC power system into devices connected to a patient. Special purpose isolation transformers may include shielding to prevent coupling of electromagnetic noise between circuits, or may have reinforced insulation to withstand thousands of volts of potential difference between primary and secondary circuits.

Instrument Transformer

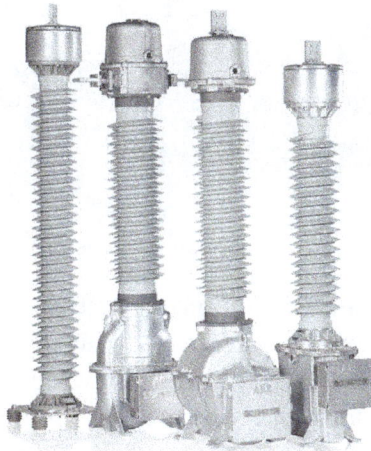

Instrument transformers

Instrument transformers are typically used to operate instruments from high voltage lines or high current circuits, safely isolating measurement and control circuitry from the high voltages or currents. The primary winding of the transformer is connected to the high voltage or high current circuit, and the meter or relay is connected to the secondary circuit. Instrument transformers may also be used as an isolation transformer so that secondary quantities may be used without affecting the primary circuitry.

Terminal identifications (either alphanumeric such as H_1, X_1, Y_1, etc. or a colored spot or dot impressed in the case) indicate one end of each winding, indicating the same instantaneous polarity and phase between windings. This applies to both types of instrument transformers. Correct identification of terminals and wiring is essential for proper operation of metering and protective relay instrumentation.

Current Transformer

A current transformer (CT) is a series connected measurement device designed to provide a current in its secondary coil proportional to the current flowing in its primary. Current transformers are commonly used in metering and protective relays in the electrical power industry.

Current transformers are often constructed by passing a single primary turn (either an insulated cable or an uninsulated bus bar) through a well-insulated toroidal core wrapped with many turns of wire. The CT is typically described by its current ratio from primary to secondary. For example, a 1000:1 CT provides an output current of 1 amperes when 1000 amperes flow through the primary winding. Standard secondary current ratings are 5 amperes or 1 ampere, compatible with standard measuring instruments. The secondary winding can be single ratio or have several tap points to provide a range of ratios. Care must be taken to make sure the secondary winding is not disconnected from its low-impedance load while current flows in the primary, as this may produce a dangerously high voltage across the open secondary and may permanently affect the accuracy of the transformer.

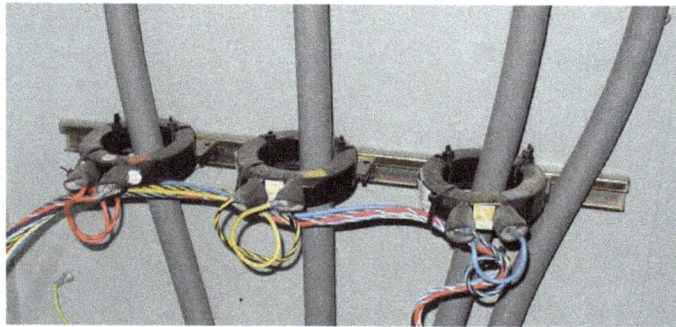

Current transformers used in metering equipment for three-phase 400 ampere electricity supply

Specially constructed wideband CTs are also used, usually with an oscilloscope, to measure high frequency waveforms or pulsed currents within pulsed power systems. One type provides a voltage output that is proportional to the measured current. Another, called a Rogowski coil, requires an external integrator in order to provide a proportional output.

A current clamp uses a current transformer with a split core that can be easily wrapped around a conductor in a circuit. This is a common method used in portable current measuring instruments but permanent installations use more economical types of current transformer.

Voltage Transformer or Potential Transformer

Voltage transformers (VT), also called potential transformers (PT), are a parallel connected type of instrument transformer, used for metering and protection in high-voltage circuits or phasor phase shift isolation. They are designed to present negligible load to the supply being measured and to have an accurate voltage ratio to enable accurate metering. A potential transformer may have several secondary windings on the same core as a primary winding, for use in different metering or protection circuits. The primary may be connected phase to ground or phase to phase. The secondary is usually grounded on one terminal.

There are three primary types of voltage transformers (VT): electromagnetic, capacitor, and optical. The electromagnetic voltage transformer is a wire-wound transformer. The capacitor voltage transformer uses a capacitance potential divider and is used at higher voltages due to a lower cost than an electromagnetic VT. An optical voltage transformer exploits the electrical properties of optical materials. Measurement of high voltages is possible by the potential transformers.

Combined Instrument Transformer

A combined instrument transformer encloses a current transformer and a voltage transformer in the same transformer. There are two main combined current and voltage transformer designs:

oil-paper insulated and SF$_6$ insulated. One advantage of applying this solution is reduced substation footprint, due to reduced number of transformers in a bay, supporting structures and connections as well as lower costs for civil works, transportation and installation.

Combined instrument transformer

Pulse Transformer

Bothhand TS6121A pulse transformer

A pulse transformer is a transformer that is optimised for transmitting rectangular electrical pulses (that is, pulses with fast rise and fall times and a relatively constant amplitude). Small versions called *signal* types are used in digital logic and telecommunications circuits, often for matching logic drivers to transmission lines. Medium-sized *power* versions are used in power-control circuits such as camera flash controllers. Larger *power* versions are used in the electrical power distribution industry to interface low-voltage control circuitry to the high-voltage gates of power semiconductors. Special high voltage pulse transformers are also used to generate high power pulses for radar, particle accelerators, or other high energy pulsed power applications.

To minimize distortion of the pulse shape, a pulse transformer needs to have low values of leakage inductance and distributed capacitance, and a high open-circuit inductance. In power-type pulse transformers, a low coupling capacitance (between the primary and secondary) is important to protect the circuitry on the primary side from high-powered transients created by the load. For the same reason, high insulation resistance and high breakdown voltage are required. A good transient response is necessary to maintain the rectangular pulse shape at the secondary, because a pulse with slow edges would create switching losses in the power semiconductors.

The product of the peak pulse voltage and the duration of the pulse (or more accurately, the voltage-time integral) is often used to characterise pulse transformers. Generally speaking, the larger this product, the larger and more expensive the transformer.

Pulse transformers by definition have a duty cycle of less than 0.5; whatever energy stored in the coil during the pulse must be "dumped" out before the pulse is fired again.

RF Transformer

There are several types of transformer used in radio frequency (RF) work. Laminated steel is not suitable for RF.

Air-core Transformer

These are used for high frequency work. The lack of a core means very low inductance. All current excites current and induces secondary voltage which is proportional to the mutual inductance. Such transformers may be nothing more than a few turns of wire soldered onto a printed circuit board.

Ferrite-core Transformer

Ferrite-core transformers are widely used in impedance matching transformers for RF, especially for baluns for TV and radio antennas. Many only have one or two turns.

Transmission-line Transformer

For radio frequency use, transformers are sometimes made from configurations of transmission line, sometimes bifilar or coaxial cable, wound around ferrite or other types of core. This style of transformer gives an extremely wide bandwidth but only a limited number of ratios (such as 1:9, 1:4 or 1:2) can be achieved with this technique.

The core material increases the inductance dramatically, thereby raising its Q factor. The cores of such transformers help improve performance at the lower frequency end of the band. RF transformers sometimes used a third coil (called a tickler winding) to inject feedback into an earlier (detector) stage in antique regenerative radio receivers.

In RF and microwave systems, a quarter-wave impedance transformer provides a way of matching impedances between circuits over a limited range of frequencies, using only a length of transmission line. The line may be coaxial cable, waveguide, stripline or microstripline.

Balun

Baluns are transformers designed specifically to connect between balanced and unbalanced circuits. These are sometimes made from configurations of transmission line and sometimes bifilar or coaxial cable and are similar to transmission line transformers in construction and operation.

IF Transformer

Ferrite-core transformers are widely used in (intermediate frequency) (IF) stages in superheterodyne radio receivers. They are mostly tuned transformers, containing a threaded ferrite slug that is screwed in or out to adjust IF tuning. The transformers are usually canned (shielded) for stability and to reduce interference.

Audio Transformer

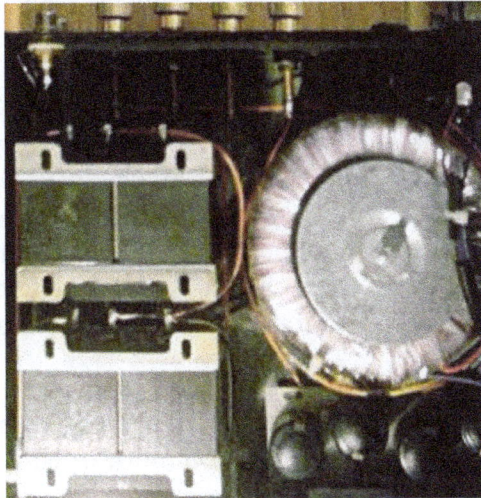

Two speaker-level audio transformers in a tube amplifier are seen on the left.
The power supply toroidal transformer is on right.

Five audio transformers for various line level purposes. The two black boxes on the left contain 1:1 transformers for splitting signals, balancing unbalanced signals, or isolating two different AC ground systems to eliminate buzz and hum. The two cylindrical metal cases fit into octal sockets; each one contains a 1:1 line transformer, the first is rated at 600 ohms, the second is rated at 15,000 ohms. On the far right is a DI unit; its 12:1 transformer (with yellow insulation) changes a high impedance unbalanced input to a low impedance balanced output.

Audio transformers are those specifically designed for use in audio circuits to carry audio signal. They can be used to block radio frequency interference or the DC component of an audio signal, to split or combine audio signals, or to provide impedance matching between high impedance and low impedance circuits, such as between a high impedance tube (valve) amplifier output and a low impedance loudspeaker, or between a high impedance instrument output and the low impedance input of a mixing console. Audio transformers that operate with loudspeaker voltages and current are larger than those that operate at microphone or line level, which carry much less power. Bridge transformers connect 2-wire and 4-wire communication circuits.

Being magnetic devices, audio transformers are susceptible to external magnetic fields such as those generated by AC current-carrying conductors. "Hum" is a term commonly used to describe unwanted signals originating from the "mains" power supply (typically 50 or 60 Hz). Audio transformers used for low-level signals, such as those from microphones, often include magnetic shielding to protect against extraneous magnetically coupled signals.

Audio transformers were originally designed to connect different telephone systems to one another while keeping their respective power supplies isolated, and are still commonly used to interconnect professional audio systems or system components, to eliminate buzz and hum. Such transformers typically have a 1:1 ratio between the primary and the secondary. These can also be used for splitting signals, balancing unbalanced signals, or feeding a balanced signal to unbalanced equipment. Transformers are also used in DI boxes to convert high-impedance instrument signals (e.g., bass guitar) to low impedance signals to enable them to connect to a microphone input on the mixing console.

A particularly critical component is the output transformer of a valve amplifier. Valve circuits for quality reproduction have long been produced with no other (inter-stage) audio transformers, but an output transformer is needed to couple the relatively high impedance (up to a few hundred ohms depending upon configuration) of the output valve(s) to the low impedance of a loudspeaker. (The valves can deliver a low current at a high voltage; the speakers require high current at low voltage.) Most solid-state power amplifiers need no output transformer at all.

Audio transformers affect the sound quality because they are non-linear. Harmonic distortion is added to the original signal, especially odd-order harmonics with an emphasis on third-order harmonics. When the incoming signal amplitude is very low there is not enough level to energize the magnetic core. When the incoming signal amplitude is very high the transformer saturates and adds ringing harmonics. Another non-linearity comes from limited frequency response. For good low-frequency response a relatively large magnetic core is required; high power handling increases the required core size. Good high-frequency response requires carefully designed and implemented windings without excessive leakage inductance or stray capacitance. All this makes for an expensive component.

Early transistor audio power amplifiers often had output transformers, but they were eliminated as advances in semiconductors allowed the design of amplifiers with sufficiently low output impedance to drive a loudspeaker directly.

Loudspeaker Transformer

In the same way that transformers create high voltage power transmission circuits that minimize transmission losses, loudspeaker transformers can power many individual loudspeakers from a single audio circuit operated at higher than normal loudspeaker voltages. This application is common in public address applications. Such circuits are commonly referred to as constant voltage speaker systems. Such systems are also known by the nominal voltage of the loudspeaker line, such as 25-, 70- and 100-volt speaker systems (the voltage corresponding to the power rating of a speaker or amplifier). A transformer steps up the output of the system's amplifer to the distribution voltage. At the distant loudspeaker locations, a step-down transformer matches the speaker to the rated voltage of the line, so the speaker produces rated nominal output when the line is at nominal voltage. Loudspeaker transformers commonly have multiple primary taps to adjust the volume at each speaker in steps.

Loudspeaker transformer in old radio

Output Transformer

Valve (tube) amplifiers almost always use an output transformer to match the high load impedance requirement of the valves (several kilohms) to a low impedance speaker.

Small Signal Transformer

Moving coil phonograph cartridges produce a very small voltage. For this to be amplified with a reasonable signal-noise ratio usually requires a transformer to convert the voltage to the range of the more common moving-magnet cartridges.

Microphones may also be matched to their load with a small transformer, which is mumetal shielded to minimise noise pickup. These transformers are less widely used today, as transistorized buffers are now cheaper.

Interstage and Coupling Transformers

In a push-pull amplifier, an inverted signal is required and can be obtained from a transformer with a center-tapped winding, used to drive two active devices in opposite phase. These phase splitting transformers are not much used today.

Other Types

Transactor

A transactor is a combination of a transformer and a reactor. A transactor has an iron core with an air-gap, which limits the coupling between windings.

Hedgehog

Hedgehog transformers are occasionally encountered in homemade 1920s radios. They are homemade audio interstage coupling transformers.

Enamelled copper wire is wound round the central half of the length of a bundle of insulated iron wire (e.g., florists' wire), to make the windings. The ends of the iron wires are then bent around the electrical winding to complete the magnetic circuit, and the whole is wrapped with tape or string to hold it together.

Variometer and Variocoupler

Variometer used in 1920s radio receiver

A variometer is a type of continuously variable air-core RF inductor with two windings. One common form consisted of a coil wound on a short hollow cylindrical form, with a second smaller coil inside, mounted on a shaft so its magnetic axis can be rotated with respect to the outer coil. The two coils are connected in series. When the two coils are collinear, with their magnetic fields pointed in the same direction, the two magnetic fields add, and the inductance is maximum. If the inner coil is rotated so its axis is at an angle to the outer coil, the magnetic fields do not add and the inductance is less. If the inner coil is rotated so it is collinear with the outer coil but their magnetic fields point in opposite directions, the fields cancel each other out and the inductance is very small or zero. The advantage of the variometer is that inductance can be adjusted continuously, over a wide range. Variometers were widely used in 1920s radio receivers. One of their main uses today is as antenna matching coils to match longwave radio transmitters to their antennas.

The *vario-coupler* was a device with similar construction, but the two coils were not connected but attached to separate circuits. So it functioned as an air-core RF transformer with variable coupling. The inner coil could be rotated from 0° to 90° angle with the outer, reducing the mutual inductance from maximum to near zero.

The pancake coil variometer was another common construction used in both 1920s receivers and transmitters. It consists of two flat spiral coils suspended vertically facing each other, hinged at one side so one could swing away from the other to an angle of 90° to reduce the coupling. The flat spiral design served to reduce parasitic capacitance and losses at radio frequencies.

Pancake or "honeycomb" coil vario-couplers were used in the 1920s in the common Armstrong or "tickler" regenerative radio receivers. One coil was connected to the detector tube's grid circuit. The other coil, the "tickler" coil was connected to the tube's plate (output) circuit. It fed back some of the signal from the plate circuit into the input again, and this positive feedback increased the tube's gain and selectivity.

Rotary Transformer

A rotary (rotatory) transformer is a specialized transformer that couples electrical signals between two parts that rotate in relation to each other—as an alternative to slip rings, which are prone to wear and contact noise. They are commonly used in helical scan magnetic tape applications.

Variable Differential Transformer

A variable differential transformer is a rugged non-contact position sensor. It has two oppositely-phased primaries which nominally produce zero output in the secondary, but any movement of the core changes the coupling to produce a signal.

Resolver and Synchro

The two-phase resolver and related three-phase synchro are rotary position sensors which work over a full 360°. The primary is rotated within two or three secondaries at different angles, and the amplitudes of the secondary signals can be decoded into an angle. Unlike variable differential transformers, the coils, and not just the core, move relative to each other, so slip rings are required to connect the primary.

Resolvers produce in-phase and quadrature components which are useful for computation. Synchros produce three-phase signals which can be connected to other synchros to rotate them in a generator/motor configuration.

Tesla coil

The Tesla coil is an electrical resonant transformer circuit designed by inventor Nikola Tesla in 1891. It is used to produce high-voltage, low-current, high frequency alternating-current electricity. Tesla experimented with a number of different configurations consisting of two, or sometimes three, coupled resonant electric circuits.

Tesla used these circuits to conduct innovative experiments in electrical lighting, phosphorescence, X-ray generation, high frequency alternating current phenomena, electrotherapy, and the transmission of electrical energy without wires. Tesla coil circuits were used commercially in sparkgap radio transmitters for wireless telegraphy until the 1920s, and in medical equipment such as electrotherapy and violet ray devices. Today their main use is for entertainment and educational displays, although small coils are still used today as leak detectors for high vacuum systems.

Operation

A Tesla coil is a radio frequency oscillator that drives an air-core double-tuned resonant transformer to produce high voltages at low currents. Tesla's original circuits as well as most modern coils use a simple spark gap to excite oscillations in the tuned transformer. More sophisticated designs use transistor or thyristor switches or vacuum tube electronic oscillators to drive the resonant transformer.

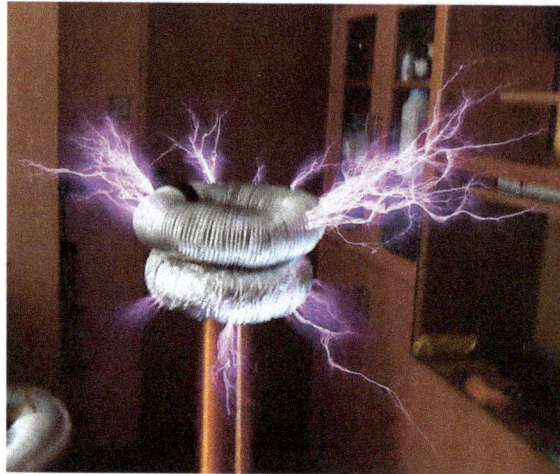

Homemade Tesla coil in operation, showing brush discharges from the toroid. The high electric field causes the air around the high voltage terminal to ionize and conduct electricity, allowing electricity to leak into the air in colorful corona discharges, brush discharges and streamer arcs. Tesla coils are used for entertainment at science museums and public events, and for special effects in movies and television.

Tesla coils can produce output voltages from 50 kilovolts to several million volts for large coils. The alternating current output is in the low radio frequency range, usually between 50 kHz and 1 MHz. Although some oscillator-driven coils generate a continuous alternating current, most Tesla coils have a pulsed output; the high voltage consists of a rapid string of pulses of radio frequency alternating current.

The common spark-excited Tesla coil circuit, shown below, consists of these components:

- A high voltage supply transformer *(T)*, to step the AC mains voltage up to a high enough voltage to jump the spark gap. Typical voltages are between 5 and 30 kilovolts (kV).

- A capacitor *(C1)* that forms a tuned circuit with the primary winding *L1* of the Tesla transformer.

- A spark gap *(SG)* that acts as a switch in the primary circuit.

- The Tesla coil *(L1, L2)*, an air-core double-tuned resonant transformer, which generates the high output voltage.

- Optionally, a capacitive electrode (top load) *(E)* in the form of a smooth metal sphere or torus attached to the secondary terminal of the coil. Its large surface area suppresses premature air breakdown and arc discharges, increasing the Q factor and output voltage.

Resonant Transformer

The specialized transformer used in the Tesla coil circuit, called a resonant transformer, oscillation transformer or radio-frequency (RF) transformer, functions differently from an ordinary transformer used in AC power circuits. While an ordinary transformer is designed to *transfer* energy efficiently from primary to secondary winding, the resonant transformer is also designed to *temporarily store* electrical energy. Each winding has a capacitance across it and functions as an LC circuit (resonant circuit, tuned circuit), storing oscillating electrical energy, analogously to a tuning fork. The primary coil *(L1)* consisting of a relatively few turns of heavy copper wire or tubing, is connected to a capacitor *(C1)* through the spark gap *(SG)*. The secondary coil *(L2)* consists of many turns (hundreds to thousands) of fine wire on a hollow cylindrical form inside the primary. The secondary is not connected to an actual capacitor, but it also functions as an LC circuit, the inductance of *(L2)* resonates with stray capacitance *(C2)*, the sum of the stray parasitic capacitance between the windings of the coil, and the capacitance of the toroidal metal electrode attached to the high voltage terminal. The primary and secondary circuits are tuned so they resonate at the same frequency, they have the same resonant frequency. This allows them to exchange energy, so the oscillating current alternates back and forth between the primary and secondary coils.

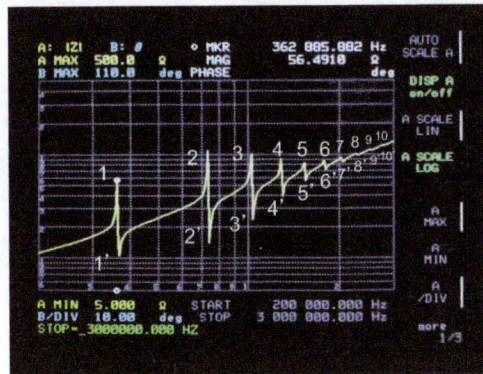

Many resonances are observed on the secondary coil

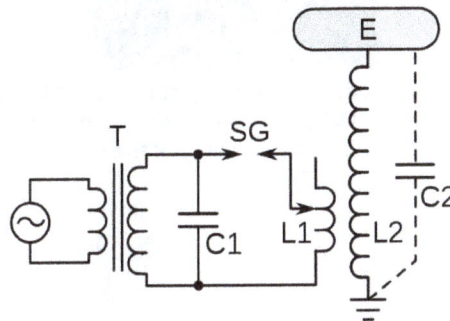

Unipolar Tesla coil circuit. *C2* is not an actual capacitor but represents the capacitance of the secondary windings *L2*, plus the capacitance to ground of the toroid electrode *E*.

The peculiar design of the coil is dictated by the need to achieve low resistive energy losses (high Q factor) at high frequencies, which results in the largest secondary voltages:

- Ordinary power transformers have an iron core to increase the magnetic coupling between the coils. However at high frequencies an iron core causes energy losses due to eddy currents and hysteresis, so it is not used in the Tesla coil.

- Ordinary transformers are designed to be "tightly coupled". Due to the iron core and close proximity of the windings, they have a high mutual inductance *(M)*, the coupling coefficient is close to unity 0.95 - 1.0, which means almost all the magnetic field of the primary winding passes through the secondary. The Tesla transformer in contrast is "loosely coupled", the primary winding is larger in diameter and spaced apart from the secondary, so the mutual inductance is lower and the coupling coefficient is only 0.05 to 0.2. This means that only 5% to 20% of the magnetic field of the primary coil passes through the secondary when it is open circuited. The loose coupling slows the exchange of energy between the primary and secondary coils, which allows the oscillating energy to stay in the secondary circuit longer before it returns to the primary and begins dissipating in the spark.

- Each winding is also limited to a single layer of wire, which reduces proximity effect losses. The primary carries very high currents. Since high frequency current mostly flows on the surface of conductors due to skin effect, it is often made of copper tubing or strip with a large surface area to reduce resistance, and its turns are spaced apart, which reduces proximity effect losses and arcing between turns.

Unipolar coil design widely used in modern coils. The primary is the flat red spiral winding at bottom, the secondary is the vertical cylindrical coil wound with fine red wire. The high voltage terminal is the aluminum torus at the top of the secondary coil.

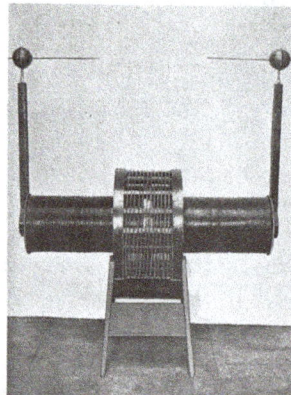

Bipolar coil, used in the early 20th century. There are two high voltage output terminals, each connected to one end of the secondary, with a spark gap between them. The primary is 12 turns of heavy wire, which is located at the midpoint of the secondary to discourage arcs between the coils.

The output circuit can have two forms:

- *Unipolar* - One end of the secondary winding is connected to a single high voltage terminal, the other end is grounded. This type is used in modern coils designed for entertainment. The primary winding is located near the bottom, low potential end of the secondary, to minimize arcs between the windings. Since the ground (Earth) serves as the return path for the high voltage, streamer arcs from the terminal tend to jump to any nearby grounded object.

- *Bipolar* - Neither end of the secondary winding is grounded, and both are brought out to high voltage terminals. The primary winding is located at the center of the secondary coil, equidistant between the two high potential terminals, to discourage arcing.

Operation Cycle

The circuit operates in a rapid, repeating cycle in which the supply transformer *(T)* charges the primary capacitor *(C1)* up, which then discharges in a spark through the spark gap, creating a brief pulse of oscillating current in the primary circuit which excites a high oscillating voltage across the secondary:

1. Current from the supply transformer *(T)* charges the capacitor *(C1)* to a high voltage.

2. When the voltage across the capacitor reaches the breakdown voltage of the spark gap *(SG)* a spark starts, reducing the spark gap resistance to a very low value. This completes the primary circuit and current from the capacitor flows through the primary coil *(L1)*. The current flows rapidly back and forth between the plates of the capacitor through the coil, generating radio frequency oscillating current in the primary circuit at the circuit's resonant frequency.

3. The oscillating magnetic field of the primary winding induces an oscillating current in the secondary winding *(L2)*, by Faraday's law of induction. Over a number of cycles, the energy in the primary circuit is transferred to the secondary. The total energy in the tuned circuits is limited to the energy originally stored in the capacitor *C1*, so as the oscillating voltage in the secondary increases in amplitude ("ring up") the oscillations in the primary decrease to zero ("ring down"). Although the ends of the secondary coil are open, it also acts as a tuned circuit due to the capacitance *(C2)*, the sum of the parasitic capacitance between the turns of the coil plus the capacitance of the toroid electrode *E*. Current flows rapidly back and forth through the secondary coil between its ends. Because of the small capacitance, the oscillating voltage across the secondary coil which appears on the output terminal is much larger than the primary voltage.

4. The secondary current creates a magnetic field that induces voltage back in the primary coil, and over a number of additional cycles the energy is transferred back to the primary. This process repeats, the energy shifting rapidly back and forth between the primary and secondary tuned circuits. The oscillating currents in the primary and secondary gradually die out ("ring down") due to energy dissipated as heat in the spark gap and resistance of the coil.

5. When the current through the spark gap is no longer sufficient to keep the air in the gap ionized, the spark stops ("quenches"), terminating the current in the primary circuit. The oscillating current in the secondary may continue for some time.

6. The current from the supply transformer begins charging the capacitor *C1* again and the cycle repeats.

This entire cycle takes place very rapidly, the oscillations dying out in a time of the order of a milli-second. Each spark across the spark gap produces a pulse of damped sinusoidal high voltage at the output terminal of the coil. Each pulse dies out before the next spark occurs, so the coil generates a string of damped waves, not a continuous sinusoidal voltage. The high voltage from the supply transformer that charges the capacitor is a 50 or 60 Hz sine wave. Depending on how the spark gap is set, usually one or two sparks occur at the peak of each half-cycle of the mains current, so there are more than a hundred sparks per second. Thus the spark at the spark gap appears continuous, as do the high voltage streamers from the top of the coil.

The supply transformer (T) secondary winding is connected across the primary tuned circuit. It might seem that the transformer would be a leakage path for the RF current, damping the oscilla-tions. However its large inductance gives it a very high impedance at the resonant frequency, so it acts as an open circuit to the oscillating current. If the supply transformer has inadequate leakage inductance, radio frequency chokes are placed in its secondary leads to block the RF current.

Oscillation Frequency

To produce the largest output voltage, the primary and secondary tuned circuits are adjusted to resonance with each other. Since the secondary circuit is usually not adjustable, this is generally done by an adjustable tap on L1 the primary coil. If the two coils were separate, the resonant fre-quencies of the primary and secondary circuits, f_1 and f_2, would be determined by the inductance and capacitance in each circuit

$$f_1 = \frac{1}{2\pi}\sqrt{\frac{1}{L_1 C_1}} \qquad f_2 = \frac{1}{2\pi}\sqrt{\frac{1}{L_2 C_2}}$$

However, because they are coupled together, the frequency at which the secondary resonates is af-fected by the primary circuit and the coupling coefficient k, and occurs at its antiresonant frequency

$$f_2' = \frac{1}{2\pi}\sqrt{\frac{1}{(1-k^2)L_2 C_2}}$$

So resonance, and the highest voltages occur when

$$\frac{1}{2\pi}\sqrt{\frac{1}{L_1 C_1}} = \frac{1}{2\pi}\sqrt{\frac{1}{(1-k^2)L_2 C_2}}$$

Thus the condition for resonance between primary and secondary is

$$L_1 C_1 = (1-k^2)L_2 C_2$$

However the Tesla transformer is very loosely coupled, and the coupling coefficient k is small, in the range 0.05 to 0.2. So the factor $\sqrt{1-k^2}$ is close to unity, 0.98 to 0.999, so the two resonant frequencies differ by 2% at most. Therefore most sources state the transformer is resonant when the resonant frequencies of primary and secondary are equal.

The resonant frequency of Tesla coils is in the low radio frequency (RF) range, usually between 50 kHz and 1 MHz. However, because of the impulsive nature of the spark they produce broadband radio noise, and without shielding can be a significant source of RFI, interfering with nearby radio and television reception.

Output Voltage

Large coil producing 3.5 meter (10 foot) streamer arcs, indicating a potential of millions of volts

In a resonant transformer the high voltage is produced by resonance; the output voltage is not proportional to the turns ratio, as in an ordinary transformer. It can be calculated approximately from conservation of energy. At the beginning of the cycle, when the spark starts, all of the energy in the primary circuit W_1 is stored in the primary capacitor C_1. If V_1 is the voltage at which the spark gap breaks down, which is usually close to the peak output voltage of the supply transformer T, this energy is

$$W_1 = \frac{1}{2} C_1 V_1^2$$

During the "ring up" this energy is transferred to the secondary circuit. Although some is lost as heat in the spark and other resistances, in modern coils, over 85% of the energy ends up in the secondary. At the peak (V_2) of the secondary sinusoidal voltage waveform, all the energy in the secondary   is stored in the capacitance C_2 between the ends of the secondary coil

$$W_2 = \frac{1}{2} C_2 V_2^2$$

Assuming no energy losses, $W_2 = W_1$. Substituting into this equation and simplifying, the peak secondary voltage is

$$V_2 = V_1 \sqrt{\frac{C_1}{C_2}} = V_1 \sqrt{\frac{L_2}{L_1}}.$$

The second formula above is derived from the first using the resonance condition $L_1 C_1 = L_2 C_2$. Since the capacitance of the secondary coil is very small compared to the primary capacitor, the primary voltage is stepped up to a high value.

It might seem that the output voltage could be increased indefinitely by reducing C_2 and L_1. However, as the output voltage increases, it reaches the point where the air next to the high voltage terminal ionizes and air discharges; coronas, brush discharges and streamer arcs, break out from the terminal. This happens when the electric field strength exceeds the dielectric strength of the air, about 30 kV per centimeter, and occurs first at sharp points and edges on the high voltage terminal. The resulting energy loss damps the oscillation, so the above lossless model is no longer accurate, and the voltage does not reach the theoretical maximum above. The output voltage of open-air Tesla coils is limited to around several million volts by air breakdown, but higher voltages can be achieved by coils immersed in pressurized tanks of insulating oil.

The Top Load or "Toroid" Electrode

Solid state DRSSTC Tesla coil with pointed wire attached to toroid to produce brush discharge

Most Tesla coil designs have a smooth spherical or toroidal shaped metal electrode on the high voltage terminal. The electrode serves as one plate of a capacitor, with the Earth as the other plate, forming the tuned circuit with the secondary winding. Although the "toroid" increases the secondary capacitance, which tends to reduce the peak voltage, its main effect is that its large diameter curved surface reduces the potential gradient (electric field) at the high voltage terminal, increasing the voltage threshold at which corona and streamer arcs form. Suppressing premature air breakdown and energy loss allows the voltage to build to higher values on the peaks of the waveform, creating longer, more spectacular streamers.

If the top electrode is large and smooth enough, the electric field at its surface may never get high enough even at the peak voltage to cause air breakdown, and air discharges will not occur. Some entertainment coils have a sharp "spark point" projecting from the torus to start discharges.

Types

The term "Tesla coil" is applied to a number of high voltage resonant transformer circuits.

Tesla coil circuits can be classified by the type of "excitation" they use, what type of circuit is used to apply current to the primary winding of the resonant transformer:

- Spark-excited or Spark Gap Tesla Coil (SGTC) - This type uses a spark gap to switch pulses of current through the primary, exciting oscillation in the transformer. This pulsed (disruptive) drive creates a pulsed high voltage output. Spark gaps have disadvantages due to the high primary currents they must handle. They produce a very loud noise while operating, noxious ozone gas, and high temperatures which often require a cooling system. The energy dissipated in the spark also reduces the Q factor and the output voltage.

  o Static spark gap - This is the most common type. It is used in most entertainment coils. An AC voltage from a high voltage supply transformer charges a capacitor, which discharges through the spark gap. The spark rate is not adjustable but is determined by the line frequency. Multiple sparks may occur on each half-cycle, so the pulses of output voltage may not be equally-spaced.

  o Static triggered spark gap - Commercial and industrial circuits often apply a DC voltage from a power supply to charge the capacitor, and use high voltage pulses generated by an oscillator applied to a triggering electrode to trigger the spark. This allows control of the spark rate and exciting voltage. Commercial spark gaps are often enclosed in an insulating gas atmosphere such as sulfur hexafluoride, reducing the length and thus the energy loss in the spark.

  o Rotary spark gap - These use a spark gap consisting of electrodes around the periphery of a wheel rotated by a motor, which create sparks when they pass by a stationary electrode. Tesla used this type on his big coils, and they are used today on large entertainment coils. The rapid separation speed of the electrodes quenches the spark quickly, allowing "first notch" quenching, making possible higher voltages. The wheel is usually driven by a synchronous motor, so the sparks are synchronized with the AC line frequency, the spark occurring at the same point on the AC waveform on each cycle, so the primary pulses are repeatable.

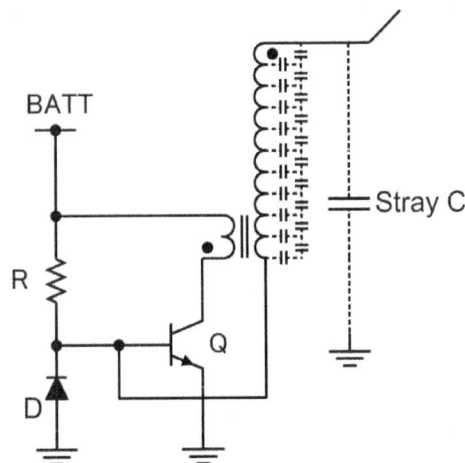

A simple single resonant solid state Tesla coil circuit in which the ground end of the secondary supplies the feedback current phase to the transistor oscillator.

- Switched or Solid State Tesla Coil (SSTC) - These use power semiconductor devices, usually thyristors or transistors such as MOSFETs or IGBTs, to switch pulses of current from a DC power supply through the primary winding. They provide pulsed (disruptive) excitation without the disadvantages of a spark gap: the loud noise, high temperatures, and poor efficiency. The voltage, frequency, and excitation waveform can be finely controllable. SSTCs are used in most commercial, industrial, and research applications as well as higher quality entertainment coils.

  - Single resonant solid state Tesla coil (SRSSTC) - In this circuit the primary does not have a capacitor and so is not a tuned circuit; only the secondary is. The pulses of current to the primary from the switching transistors excite resonance in the secondary tuned circuit. Single tuned SSTCs are simpler, but don't have as high a Q and cannot produce as high voltage from a given input power as the DRSSTC.

  - Dual Resonant Solid State Tesla Coil (DRSSTC) - The circuit is similar to the double tuned spark excited circuit, except in place of the spark gap semiconductor switches are used. This functions similarly to the double tuned spark-excited circuit. Since both primary and secondary are resonant it has higher Q and can generate higher voltage for a given input power than the SRSSTC.

  - Singing Tesla coil or musical Tesla coil - This is a Tesla coil which can be played like a musical instrument, with its high voltage discharges reproducing simple musical tones. The drive current pulses applied to the primary are modulated at an audio rate by a solid state "interrupter" circuit, causing the arc discharge from the high voltage terminal to emit sounds. Only tones and simple chords have been produced so far; the coil cannot function as a loudspeaker, reproducing complex music or voice sounds. The sound output is controlled by a keyboard or MIDI file applied to the circuit through a MIDI interface. Two modulation techniques have been used: AM (amplitude modulation of the exciting voltage) and PFM (pulse-frequency modulation). These are mainly built as novelties for entertainment.

- Continuous wave - In these the transformer is driven by a feedback oscillator, which applies a sinusoidal current to the transformer. The primary tuned circuit serves as the tank circuit of the oscillator, and the circuit resembles a radio transmitter. Unlike the previous circuits which generate a pulsed output, they generate a continuous sine wave output. Power vacuum tubes are often used as active devices instead of transistors because they are more robust and tolerant of overloads. In general, continuous excitation produces lower output voltages from a given input power than pulsed excitation.

Tesla circuits can also be classified by how many coils (inductors) they contain:

- Two coil or double-resonant circuits - Virtually all present Tesla coils use the two coil resonant transformer, consisting of a primary winding to which current pulses are applied, and a secondary winding that produces the high voltage, invented by Tesla in 1891. The term "Tesla coil" normally refers to these circuits.

- Three coil, triple-resonant, or magnifier circuits - These are circuits with three coils, based on Tesla's "magnifying transmitter" circuit which he began experimenting with sometime before 1898 and installed in his Colorado Springs lab 1899-1900, and patented in 1902. They consist of a two coil air-core step-up transformer similar to the Tesla transformer, with the secondary connected to a third coil not magnetically coupled to the others, called the "extra" or "resonator" coil, which is series-fed and resonates with its own capacitance. The presence of three energy-storing tank circuits gives this circuit more complicated resonant behavior. It is the subject of research, but has been used in few practical applications.

History

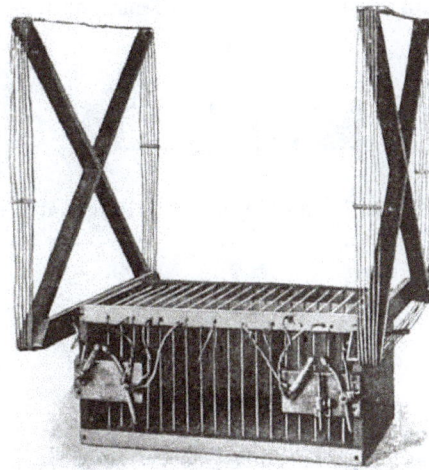

Henry Rowland's 1889 spark-excited resonant transformer, a predecessor to the Tesla coil

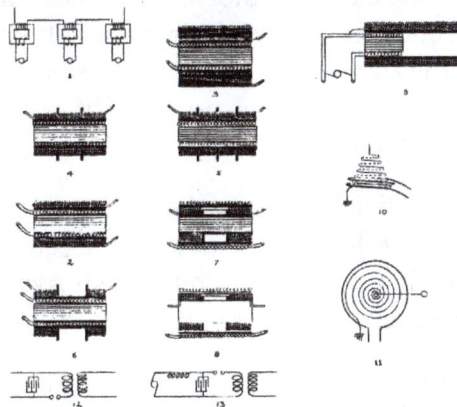

Steps in Tesla's development of the Tesla transformer around 1891. (1) Closed-core transformers used at low frequencies, (2-7) rearranging windings for lower losses, (8) removed iron core, (9) partial core, (10-11) final conical Tesla transformer, (12-13) Tesla coil circuits.

Nikola Tesla patented the Tesla coil circuit April 25, 1891. and first publicly demonstrated it May 20, 1891 in his lecture *"Experiments with Alternate Currents of Very High Frequency and Their Application to Methods of Artificial Illumination"* before the American Institute of Electrical Engineers at Columbia College, New York. Although Tesla patented many similar circuits during this period, this was the first that contained all the elements of the Tesla coil: high voltage primary transformer, capacitor, spark gap, and air core "oscillation transformer".

Invention

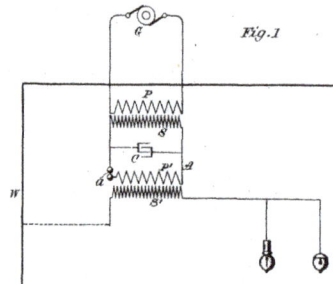

First drawing of Tesla coil circuit from Tesla's April 25, 1891 patent

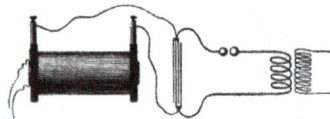

Drawing of Tesla coil circuit from Tesla's May 20, 1891 lecture at Columbia College, New York

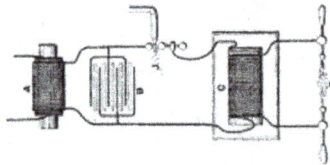

Elihu Thomson's Tesla coil, published February 1892, identical to Tesla's except for a compressed air spark blowout *(J)*.

During the Industrial Revolution the electrical industry exploited direct current (DC) and low frequency alternating current (AC), but not much was known about frequencies above 20 kHz, what are now called radio frequencies. In 1887, four years previously, Heinrich Hertz had discovered Hertzian waves (radio waves), electromagnetic waves which oscillated at very high frequencies. This attracted much attention, and a number of researchers began experimenting with high frequency currents.

Tesla's background was in the new field of alternating current power systems, so he understood transformers and resonance. In 1888 he decided that high frequencies were the most promising field for research, and set up a laboratory at 33 South Fifth Avenue, New York for researching them, initially repeating Hertz's experiments.

He first developed alternators as sources of high frequency current, but by 1890 found they were limited to frequencies of about 20 kHz. In search of higher frequencies he turned to spark-excited resonant circuits. Tesla's innovation was in applying resonance to transformers. Transformers functioned differently at high frequencies than at the low frequencies used in power systems; the iron core in low frequency transformers caused energy losses due to eddy currents and hysteresis. Tesla and Elihu Thomson independently developed a new type of transformer without an iron core, the "oscillation transformer", and the Tesla coil circuit to drive it to produce high voltages.

Tesla invented the Tesla coil during efforts to develop a "wireless" lighting system, with gas discharge light bulbs that would glow in an oscillating electric field from a high voltage, high frequency power source. For a high frequency source Tesla powered a Ruhmkorff coil (induction coil)

with his high frequency alternator. He found that the core losses due to the high frequency current overheated the iron core in the Ruhmkorff coil and melted the insulation between the primary and secondary windings. To fix this problem Tesla changed the design so that there was an air gap instead of insulating material between the windings, and made the iron core adjustable so it could be moved in or out of the coil He eventually found the highest voltages could be produced when the iron core was omitted. Tesla also found he needed to put the capacitor normally used in the Ruhmkorff circuit between his alternator and the coil's primary winding to avoid burning out the coil. By adjusting the coil and capacitor Tesla found he could take advantage of the resonance set up between the two to achieve even higher frequencies. He found that the highest voltages were generated when the "closed" primary circuit with the capacitor was in resonance with the "open" secondary winding.

One of Tesla's early coils at his New York lab in 1892, with a conical secondary

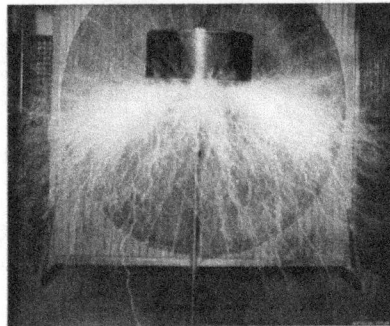

Prototype "magnifying transmitter" in Tesla's New York lab around 1898 producing 2.5 million volts. The round "spiderweb" secondary coil is visible in background.

Compact coil designed by Tesla for use as an ozone generator for water treatment

Tesla was not the first to invent this circuit. Henry Rowland built a spark-excited resonant transformer circuit *(above)* in 1889 and Elihu Thomson had experimented with similar circuits in 1890, including one which could produce 64 inch (1.6 m) sparks, and other sources confirm Tesla was not the first. However he was the first to see practical applications for it and patent it. Tesla did not perform detailed mathematical analyses of the circuit, relying instead on trial and error and his intuitive understanding of resonance. He even realized that the secondary coil functioned as a quarter-wave resonator; he specified the length of the wire in the secondary coil must be a quarter wavelength at the resonant frequency. The first mathematical analyses of the circuit were done by Anton Oberbeck (1895) and Paul Drude (1904).

Tesla's Demonstrations

Tesla demonstrating wireless lighting at his 1891 lecture at Columbia College. The two metal sheets are connected to a Tesla coil oscillator, which applies a high radio frequency oscillating voltage. The oscillating electric field between the sheets ionizes the low pressure gas in the two long Geissler tubes he is holding, causing them to glow by fluorescence, similar to neon lights, without wires.

A charismatic showman and self-promoter, in 1891-1893 Tesla used the Tesla coil in dramatic public lectures demonstrating the new science of high voltage, high frequency electricity. The radio frequency AC electric currents produced by a Tesla coil did not behave like the DC or low frequency AC current scientists of the time were familiar with. In lectures at Columbia College May 20, 1891, scientific societies in Britain and France during a 1892 European speaking tour, the Franklin Institute, Philadelphia in February 1893, and the National Electric Light Association, St. Louis in March 1893, he impressed audiences with spectacular brush discharges and streamers, heated iron by induction heating, showed RF current could pass through insulators and be conducted by a single wire without a return path, and powered light bulbs and motors without wires. He demonstrated that high frequency currents often did not cause the sensation of electric shock, applying hundreds of thousands of volts to his own body, causing his body to light up with a glowing corona discharge in the darkened room. These lectures introduced the "Tesla oscillator" to the scientific community, and made Tesla internationally famous.

Wireless Power Experiments

Tesla employed the Tesla coil in his efforts to achieve wireless power transmission, his lifelong dream. In the period 1891 to 1900 he used it to perform some of the first experiments in wireless power, transmitting radio frequency power across short distances by inductive coupling between coils of wire. In his early 1890s demonstrations such as those before the American Institute of Electrical Engineers and at the 1893 Columbian Exposition in Chicago he lit light bulbs from across a room. He found he could increase the distance by using a receiving LC circuit tuned to resonance with the Tesla coil's LC circuit, transferring energy by resonant inductive coupling. At his Colorado Springs laboratory during 1899-1900, by using voltages of the order of 10 million volts generated by his enormous magnifying transmitter coil, he was able to light three incandescent lamps at a distance of about 100 feet (30 m). Today the resonant inductive coupling discovered by Tesla is a familiar concept in electronics, widely used in IF transformers and short range wireless power transmission systems such as cellphone charging pads.

Light bulb *(bottom)* powered wirelessly by "receiver" coil tuned to resonance with the huge "magnifying transmitter" coil at Tesla's Colorado Springs lab, 1899.

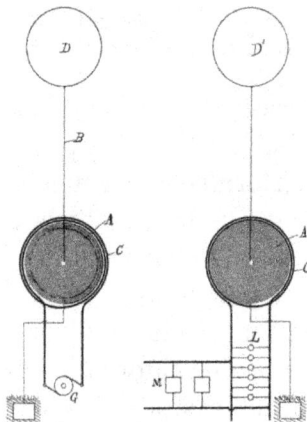

Tesla's proposed wireless power system, from his 1897 patent. The transmitter *(left)* consists of a Tesla coil *(A,C)* driving an elevated capacitive terminal *(B)* suspended by a balloon *(D)*. The receiver *(right)* is a similar terminal and resonant transformer.

It is now understood that inductive and capacitive coupling are "near-field" effects, so they cannot be used for long-distance transmission. However, Tesla was obsessed with developing a long range wireless power transmission system which could transmit power from power plants directly into homes and factories without wires, described in a visionary June 1900 article in *Century Magazine*; "The Problem of Increasing Human Energy". He claimed to be able to transmit power on a *worldwide* scale, using a method that involved conduction through the Earth and atmosphere. Tesla believed that the entire Earth could act as an electrical resonator, and that by driving current pulses into the Earth at its resonant frequency from a grounded Tesla coil with an elevated capacitance, the potential of the Earth could be made to oscillate, creating global standing waves, and this alternating current could be received with a capacitive antenna tuned to resonance with it at any point on Earth. Another of his ideas was that transmitting and receiving terminals could be suspended in the air by balloons at 30,000 feet (9,100 m) altitude, where the air pressure is lower. At this altitude, he thought, a layer of electrically conductive rarefied air would allow electricity to be sent at high voltages (hundreds of millions of volts) over long distances. Tesla envisioned building a global network of wireless power stations, which he called his "World Wireless System", which would transmit both information and electric power to everyone on Earth. There is no reliable evidence that he ever transmitted significant amounts of power beyond the short range.

Magnifying Transmitter

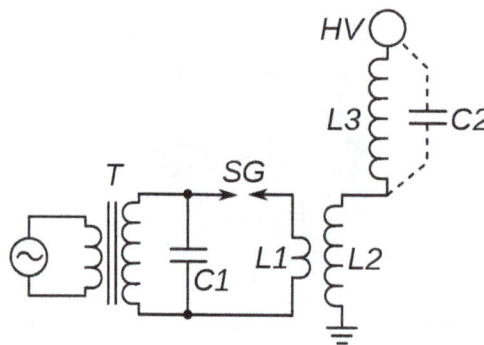

Circuit of magnifying transmitter at Tesla's Colorado Springs laboratory. *C2* represents the parasitic capacitance between the windings of coil *L3*.

Tesla's wireless research required increasingly high voltages, and he had reached the limit of the voltages he could generate within the space of his New York lab. Between 1899-1900 he built a laboratory in Colorado Springs and performed experiments on wireless transmission there. The Colorado Springs laboratory had one of the largest Tesla coils ever built, which Tesla called a "magnifying transmitter" as it was intended to transmit power to a distant receiver. With an input power of 300 kilowatts it could produce potentials of the order of 10 million volts, at frequencies of 50–150 kHz, creating huge "lightning bolts" reportedly up to 135 feet long. During experiments, it caused an overload which set fire to the alternator of the Colorado Springs power company, destroying it, and Tesla had to rebuild the alternator.

In the magnifying transmitter, Tesla used a modified design which he had been experimenting with since before 1898 and patented in 1902, different from his previous double-tuned circuits. In addition to the primary (*L1*) and secondary (*L2*) coils, it had a third coil (*L3*) which he called the "extra" coil, not magnetically coupled to the others, attached to the top terminal of the secondary. When driven by the secondary it produced additional high voltage by resonance, being adjusted to resonate with its own parasitic capacitance (*C2*) The use of a series-fed resonator coil to generate high voltages was independently discovered by Paul Marie Oudin in 1893 and employed in his Oudin coil.

The Colorado Springs apparatus consisted of a 51 foot (15.5 m) diameter Tesla transformer composed of a secondary winding *(L2)* of 50 turns of heavy wire wound on an 6 foot (2 m) high circular wooden "fence" around the periphery of the lab, and a single-turn primary *(L1)* either mounted on the fence or buried in the ground under it. The primary was connected to a bank of oil capacitors *(C1)* to make a tuned circuit, with a rotary spark gap *(SG)*, powered by 20 to 40 kilovolts from a powerful utility step-up transformer *(T)*. The top of the secondary was connected to the 100-turn 8 ft (2.4 m) diameter "extra" or "resonator" coil *(L3)* in the center of the room. It's high voltage end was connected to a telescoping 143 foot (43.6 m) "antenna" rod with a 30 inch (1 m) metal ball on top which could project through the roof of the lab. By cranking the rod up or down he could adjust the capacitance in the circuit of the extra coil, tuning it to resonance with the rest of the circuit.

Wardenclyffe Tower

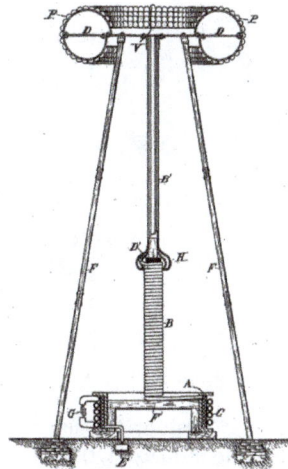

Design on which the Wardenclyffe plant was based, from Tesla's 1902 patent

In 1901, convinced his wireless theories were correct, Tesla with financing from banker J. P. Morgan began construction of a high-voltage wireless station, now called the Wardenclyffe Tower, at Shoreham, New York. Although it was built as a transatlantic radiotelegraphy station, Tesla also intended it to transmit electric power without wires as a prototype transmitter for his proposed "World Wireless System". Essentially an enormous Tesla coil, it consisted of a powerhouse with a 400 horsepower generator and a 187 foot (57 m) tower topped by a 68 foot (21 m) diameter metal dome capacitive electrode. Underneath the surface was an elaborate ground system that Tesla said was needed to "grip the earth" to create the oscillating earth currents which he believed would transmit the power.

By 1904 his investors had pulled out and the facility was never completed; it was torn down in 1916. Although Tesla seems to have believed his wireless power ideas were proven, he had a history of making claims that he had not confirmed by experiment, and there seems to be no evidence that he ever transmitted significant power beyond the short-range demonstrations mentioned above. The few reports of long-distance power transmission by Tesla are not from reliable sources. For example, a widely repeated myth is that in 1899 he wirelessly lit 200 light bulbs at a distance of 26 miles (42 km). There is no independent confirmation of this supposed demonstration; Tesla did not mention it, and it does not appear in his laboratory notes. It originated in 1944 from Tesla's first biographer, John J. O'Neill, who said he pieced it together from "fragmentary material... in a number of publications".

In the 100 years since, others such as Robert Golka have built equipment similar to Tesla's, but long distance power transmission has not been demonstrated, and the scientific consensus is his World Wireless system would not have worked. Contemporary scientists point out that while Tesla's coils (with appropriate antennas) can function as radio transmitters, transmitting energy in the form of radio waves, the frequency he used, around 150 kHz, is far too low for practical long range power transmission. At these wavelengths the radio waves spread out in all directions and cannot be focused on a distant receiver. Tesla's world power transmission scheme remains today what it was in Tesla's time: a bold, fascinating dream.

Use in Radio

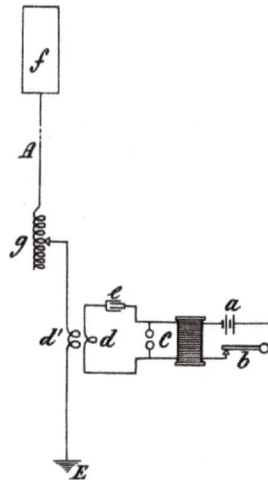

Spark transmitter circuit from Marconi's 1900 patent. It's similarity to a Tesla coil can be seen; the only difference is the addition of a variable inductor (*g*) to tune the antenna (*f*) to resonance.

"[The Tesla coil] *was invented not for wireless but for making vacuum lamps glow without external electrodes, and it later played a principal part in other hands in the operation of big spark stations.*"

--William H. Eccles, 1933

One of the largest applications of the Tesla coil circuit was in early radio transmitters called spark gap transmitters. The first radio wave generators, invented by Heinrich Hertz in 1887, were spark gaps connected directly to antennas, powered by induction coils. Because they lacked a resonant circuit, these transmitters produced highly damped radio waves. As a result their transmissions occupied an extremely wide bandwidth of frequencies. When multiple transmitters were operating in the same area their frequencies overlapped and they interfered with one another, causing garbled reception. There was no way for a receiver to select one signal over another.

In 1892 William Crookes, a friend of Tesla, had given a lecture on the uses of radio waves in which he suggested using resonance to reduce the bandwidth in transmitters and receivers. By using resonant circuits, different transmitters could be "tuned" to transmit on different frequencies. With narrower bandwidth, separate transmitter frequencies would no longer overlap, so a receiver could receive a particular transmission by "tuning" its resonant circuit to the same frequency as the transmitter. This is the system used in all modern radio.

With an appropriate wire antenna, the Tesla coil circuit could function as such a narrow band-width radio transmitter. In his March 1893 St. Louis lecture, Tesla demonstrated a wireless system that was the first use of tuned circuits in radio, although he used it for wireless power transmission, not radio communication. A grounded spark-excited capacitor-tuned Tesla transformer attached to an elevated wire antenna transmitted radio waves, which were received across the room by a wire antenna attached to a receiver consisting of a second grounded resonant transformer tuned to the transmitter's frequency, which lighted a Geissler tube. This system, patented by Tesla September 2, 1897, was the first use of the "four circuit" concept later claimed by Marconi. However, Tesla was mainly interested in wireless power and never developed a practical radio *communication* system. In fact, he never believed that radio waves could be used for practical communication, instead clinging to an erroneous theory that radio communication was due to currents in the Earth.

Practical radiotelegraphy communication systems were developed by Guglielmo Marconi beginning in 1895. By 1897 the advantages of narrow-bandwidth (lightly damped) systems noted by Crookes were recognized, and resonant circuits, capacitors and inductors, were incorporated in transmitters and receivers. The "closed primary, open secondary" resonant transformer circuit used by Tesla proved a superior transmitter, because the loosely-coupled transformer partially isolated the oscillating primary circuit from the energy-radiating antenna circuit, reducing the damping, allowing it to produce long "ringing" waves which had a narrower bandwidth. Versions of the circuit were patented by Marconi, John Stone Stone and Oliver Lodge, and were widely used in radio for twenty years. In 1906 Max Wien invented the quenched or "series" spark gap, which extinguished the spark after the energy had been transferred to the secondary, allowing the secondary to oscillate freely after that, reducing damping and bandwidth still more.

Although their damping had been reduced as much as possible, spark transmitters still produced damped waves which had a wide bandwidth, creating interference with other transmitters. Around 1920 they became obsolete, superseded by vacuum tube transmitters which generated continuous waves at a single frequency, which could also be modulated to carry sound. Tesla's resonant transformer continued to be used in vacuum tube transmitters and receivers, and is a key component in radio to this day.

During the "spark era" the radio engineering profession gave credit to Tesla, his circuit became known as the "Tesla coil" or "Tesla transformer". However Tesla did not benefit financially, due to competing patent claims. Marconi had claimed rights to the "closed primary open secondary" transmitter circuit in his controversial 1900 "four circuit" wireless patent. Tesla sued Marconi in 1915 for patent infringement, but didn't have the resources to pursue the action. However in 1943, in a separate suit brought by the Marconi Company against the US government for use of its patents in WW1, the US Supreme Court invalidated Marconi's 1900 patent claim to the "four circuit" concept. The ruling cited the prior patents of Tesla, Lodge, and Stone, but did not decide which of these parties had rights to the circuit. Of course by this time the issue was moot; the patent had expired in 1915 and spark transmitters had long been obsolete.

Although there is some disagreement over the role Tesla himself played in the invention of radio, sources agree on the importance of his circuit in early radio transmitters. From a modern perspective, most spark transmitters could be regarded as Tesla coils.

Use in Medicine

Vacuum electrode "violet ray" wand in operation

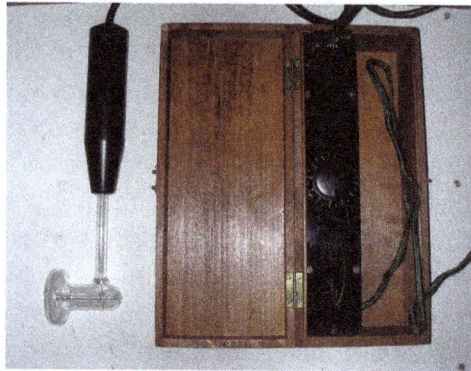

A violet ray wand, a handheld Tesla coil sold as a quack home medical device until about 1940. Said to cure everything from carbuncles to lumbago.

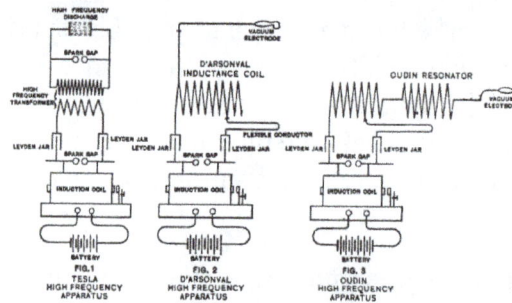

The three circuits used in electrotherapy apparatus in the early 20th century: (1) Tesla coil, (2) D'Arsonval coil, (3) Oudin coil. In medical coils for safety *two* capacitors (Leyden jars) were used, one in each branch of the primary circuit, to completely isolate the patient's body from the potentially lethal currents of the supply transformer, in case of an electrical fault.

Tesla had observed as early as 1891 that high frequency currents above 10 kHz did not cause the sensation of electric shock, and in fact currents that would be lethal at lower frequencies could be passed through the body without apparent harm. He experimented on himself, and claimed daily applications of high voltage relieved depression. He was one of the first to observe the heating effect of high frequency currents on the body, the basis of diathermy. During his highly-publicized early 1890s demonstrations he passed hundreds of thousands of volts through his body. With characteristic hyperbole he called electricity "the greatest of all doctors" and suggested burying wires under classrooms so its stimulating effect would improve performance of "dull" schoolchildren. Tesla wrote a pioneering paper in 1898 on the medical uses of high frequency currents but did little further work on the subject.

A few other researchers were also experimentally applying high frequency currents to the body at this time. Elihu Thomson, the co-inventor of the Tesla coil, was one, so in medicine the Tesla coil became known as the "Tesla-Thomson apparatus". In France, from 1889 physician and pioneering biophysicist Jacques d'Arsonval had been documenting the physiological effects of high frequency current on the body, and had made the same discoveries as Tesla. During his 1892 European trip Tesla met with D'Arsonval and was flattered to find they were using similar circuits. D'Arsonval's spark-excited resonant circuits (above) did not produce as high voltage as the Tesla transformer. In 1893 French physician Paul Marie Oudin added a "resonator" coil to the D'Arsonval circuit to create the high voltage Oudin coil, a circuit very similar to the Tesla coil, which was widely used for treating patients in Europe.

During this period, people were fascinated by the new technology of electricity, and many believed it had miraculous curative or "vitalizing" powers. Medical ethics were also looser, and doctors could experiment on their patients. By the turn of the century, application of high voltage, "high frequency" currents to the body had become part of a Victorian era medical field, part legitimate experimental medicine and part quack medicine, called *electrotherapy*. Manufacturers produced medical apparatus to generate "Tesla currents", "D'Arsonval currents", and "Oudin currents" for physicians. In electrotherapy, a pointed electrode attached to the high voltage terminal of the coil was held near the patient, and the luminous brush discharges from it (called "*effluves*") were applied to parts of the body to treat a wide variety of medical conditions. In order to apply the electrode directly to the skin, or tissues inside the mouth, anus or vagina, a "vacuum electrode" was used, consisting of a metal electrode sealed inside a partially evacuated glass tube, which produced a dramatic violet glow. The glass wall of the tube and the skin surface formed a capacitor which limited the current to the patient, preventing discomfort. These vacuum electrodes were later manufactured with handheld Tesla coils to make "violet ray" wands, sold to the public as a quack home medical device.

The popularity of electrotherapy peaked after World War 1, but by the 1920s authorities began to crack down on fraudulent medical treatments, and electrotherapy largely became obsolete. A part of the field that survived was *diathermy*, the application of high frequency current to heat body tissue, pioneered by German physician Karl Nagelschmidt in 1907 using Tesla coils. By 1930 "long wave" (0.5~2 MHz) Tesla coil diathermy machines were being replaced by "short wave" (10~100 MHz) vacuum tube diathermy machines, but Tesla coils continued to be used in both diathermy and quack medical devices like violet ray until World War 2.

During the 1920s and 30s all unipolar (single terminal) high voltage medical coils came to be called Oudin coils, so today's unipolar Tesla coils are sometimes referred to as "Oudin coils".

Use in Show Business

The Tesla coil's spectacular displays of sparks, and the fact that its currents could pass through the human body without causing electric shock, led to its use in the entertainment business.

In the early 20th century it appeared in traveling carnivals, freak shows and circus and carnival sideshows, which often had an act in which a performer would pass high voltages through his body Performers such as "Dr. Resisto", "The Human Dynamo", "Electrice", "The Great Volta", and

"Madamoiselle Electra" would have their body connected to the high voltage terminal of a hidden Tesla coil, causing sparks to shoot from their fingertips and other parts of their body, and Geissler tubes to light up when held in their hand or even brought near them. They could also light candles or cigarettes with their fingers. Although they didn't usually cause electric shocks, RF arc discharges from the bare skin could cause painful burns; to prevent them performers sometimes wore metal thimbles on their fingertips *(Rev. Moon, center image above, is using them)*. These acts were extremely dangerous and could kill the performer if the Tesla coil was misadjusted. In carny lingo this was called an "electric chair act" because it often included a spark-laced "electrocution" of the performer in an electric chair, exploiting public fascination with this exotic new method of capital punishment, which had become the United States' dominant method of execution around 1900. Today entertainers still perform high voltage acts with Tesla coils, but modern bioelectromagnetics has brought a new awareness of the hazards of Tesla coil currents, and allowing them to pass through the body is today considered extremely dangerous.

Tesla coils were also used as dramatic props in early mystery and science fiction motion pictures, starting in the silent era. The crackling, writhing sparks emanating from the electrode of a giant Tesla coil became Hollywood's iconic symbol of the "mad scientist's" lab, recognized throughout the world. This was probably because the eccentric Nikola Tesla himself, with his famous high voltage demonstrations and his mysterious Colorado Springs laboratory, was one of the main prototypes from which the "mad scientist" stock character originated. Some early films in which Tesla coils appeared were *Wolves of Kultur* (1918), *The Power God* (1926), *Metropolis* (1927), *Frankenstein* (1931) and its many sequels such as *Son of Frankenstein* (1939), *The Mask of Fu Manchu* (1932), *Chandu the Magician* (1932), *The Lost City* (1935), and *The Clutching Hand* (1936) and many later films and television shows. By the 1980s, effects like high voltage sparks were being added to movies by CGI as visual effects in post-production, eliminating the need for dangerous high voltage Tesla coils on sets.

The Tesla coils for many of these movies were constructed by Kenneth Strickfaden (1896-1984) who, beginning with his spectacular effects in the 1931 *Frankenstein*, became Hollywood's preeminent electrical special effects expert. His large "Meg Senior" Tesla coil seen in many of these movies consisted of a 6 foot 1000 turn conical secondary and a 10 turn primary, connected to a capacitor through a rotary spark gap, powered by a 20 kV transformer. It could produce 6 foot sparks. Some of his last gigs were the reassembly of the original 1931 *Frankenstein* high voltage apparatus for the Mel Brooks satire *Young Frankenstein* (1974), and construction of a million volt Tesla coil which produced 12 foot sparks for a 1976 stage show by the rock band *Kiss*.

Use in Education

Small educational Tesla coil kit, 1918

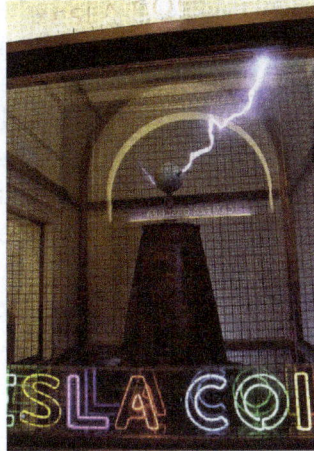

Million volt Griffith Park Observatory coil, Los Angeles. Over 100 years old,
it is one of the oldest working Tesla coils.

Demonstration of inductance with a Tesla coil, 1906. RF current will not pass through the
heavy copper wire because of the bend, and passes through the lamp instead.

Ever since Tesla's 1890s lectures, Tesla coils have been used as attractions in educational exhibits and science fairs. They have become a way to counter the stereotype that science is boring. In the early 20th century, experts like Henry Transtrom and Earle Ovington gave high voltage demonstrations at "electric fairs". High school classes built Tesla coils.

From 1933 into the 1980s, between movie jobs Hollywood special effects expert Ken Strickfaden would take his high voltage apparatus on the road in an exhibition called "Science on Parade" and later "The Kenstric Space Age Science Show" to high schools, colleges, World Fairs and expositions. These spectacular shows, which reached 48 states, had a seminal influence on the birth of the modern "coiling" movement. A number of present-day Tesla hobbyists such as William Wysock say they were inspired to build Tesla coils by seeing Strickfaden's show.

One of the oldest and best-known coils still in operation is the "GPO-1" at Griffith Park Observatory in Los Angeles. It was originally one of a pair of coils built in 1910 by Earle L. Ovington, a friend of Tesla and manufacturer of high voltage electrotherapy apparatus. For a number of years Ovington displayed them at the December electrical trade show at Madison Square Garden in New York City, using them for demonstrations of high voltage science, which Tesla himself sometimes attended. Called the *Million Volt Oscillator*, the twin coils were installed on the balcony at the show. Every hour the lights were dimmed and the public was treated to a display of 10 foot arcs. Ovington gave the coils to his friend Dr. Frederick Finch Strong, a leading figure in the alternative health field of electrotherapy. In 1937 Strong donated the coils to the Griffith Observatory. The museum didn't have room to display both, but one coil was restored by Kenneth Strickfaden and has been in daily operation ever since. It consists of a 48 in. (1.2 m) high conical secondary coil topped by a 12 in. (30 cm) diameter copper ball electrode, with a 9-turn spiral primary of 2 in. copper strip, a glass plate capacitor (replacing the original Leyden jars), and rotary spark gap. Its output has been estimated at 1.3 million volts.

Later uses

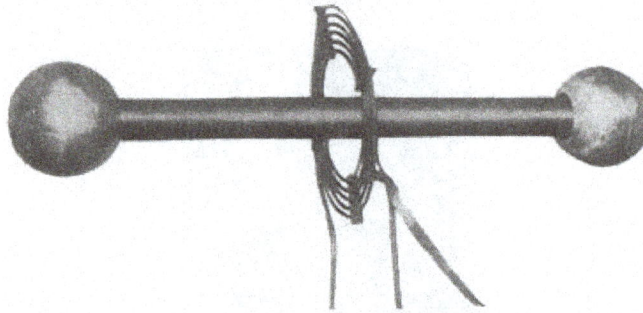

Breit and Tuve's 5 MV Tesla coil used as particle accelerator, 1928

In addition to its use in spark-gap radio transmitters and electrotherapy described above, the Tesla coil circuit was also used in the early 20th century in x-ray machines, ozone generators for water purification, and induction heating equipment. However in the 1920s vacuum tube oscillators replaced it in all these applications. The triode vacuum tube was a much better radio frequency current generator than the noisy, hot, ozone-producing spark, and could produce continuous waves. After this, industrial use of the Tesla coil was mainly limited to a few specialized applications which were suited to its unique characteristics, such as high voltage insulation testing.

In 1926, pioneering accelerator physicists Merle Tuve and Gregory Breit built a 5 million volt Tesla coil as a linear particle accelerator. The bipolar coil consisted of a pyrex tube a meter long wound with 8000 turns of fine wire, with round corona caps on each end, and a 5 turn spiral primary coil surrounding it at the center. It was operated in a tank of insulating oil pressurized to 500 psi which allowed it to reach a potential of 5.2 megavolts. Although it was used for a short period in 1929-30 it was not a success because the particles' acceleration had to be completed within the brief period of a half cycle of the RF voltage.

In 1970 Robert K. Golka built a replica of Tesla's huge Colorado Springs magnifying transmitter in a shed at Wendover Air Force Base, Utah, using data he found in Tesla's lab notes archived at the Nikola Tesla Museum in Beograd, Serbia. This was one of the first experiments with the magnifier circuit since Tesla's time. The coil generated 12 million volts. Golka used it to try to duplicate Tesla's reported synthesis of ball lightning.

Modern-day Tesla Coils

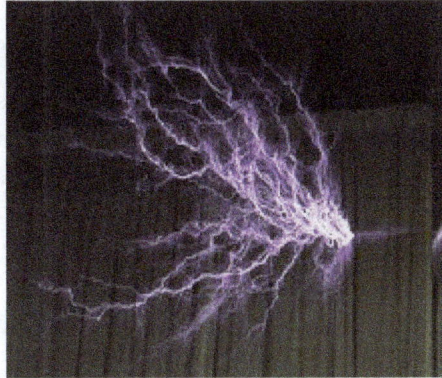

Electric discharge showing the lightning-like plasma filaments from a 'Tesla coil'

Tesla coil (discharge)

Tesla coil in terrarium (I)

Modern high-voltage enthusiasts usually build Tesla coils similar to some of Tesla's "later" 2-coil air-core designs. These typically consist of a primary tank circuit, a series LC (inductance-capacitance) circuit composed of a high-voltage capacitor, spark gap and primary coil, and the secondary LC circuit, a series-resonant circuit consisting of the secondary coil plus a terminal capacitance or "top load". In Tesla's more advanced (magnifier) design, a third coil is added. The secondary LC circuit is composed of a tightly coupled air-core transformer secondary coil driving the bottom of a separate third coil helical resonator. Modern 2-coil systems use a single secondary coil. The top of the secondary is then connected to a topload terminal, which forms one 'plate' of a capacitor, the other 'plate' being the earth (or "ground"). The primary LC circuit is tuned so that it resonates at the same frequency as the secondary LC circuit. The primary and secondary coils are magnetically coupled, creating a dual-tuned resonant air-core transformer. Earlier oil-insulated Tesla coils needed large and long insulators at their high-voltage terminals to prevent discharge in air. Later Tesla coils spread their electric fields over larger distances to prevent high electrical stresses in the first place, thereby allowing operation in free air. Most modern Tesla coils also use toroid-shaped output terminals. These are often fabricated from spun metal or flexible aluminum ducting. The toroidal shape helps to control the high electrical field near the top of the secondary by directing sparks outward and away from the primary and secondary windings.

A more complex version of a Tesla coil, termed a "magnifier" by Tesla, uses a more tightly coupled air-core resonance "driver" transformer (or "master oscillator") and a smaller, remotely located output coil (called the "extra coil" or simply the resonator) that has a large number of turns on a relatively small coil form. The bottom of the driver's secondary winding is connected to ground. The opposite end is connected to the bottom of the extra coil through an insulated conductor that is sometimes called the transmission line. Since the transmission line operates at relatively high RF voltages, it is typically made of 1" diameter metal tubing to reduce corona losses. Since the third coil is located some distance away from the driver, it is not magnetically coupled to it. RF energy is instead directly coupled from the output of the driver into the bottom of the third coil, causing it to "ring up" to very high voltages. The combination of the two-coil driver and third coil resonator adds another degree of freedom to the system, making tuning considerably more complex than that of a 2-coil system. The transient response for multiple resonance networks (of which the Tesla magnifier is a sub-set) has only recently been solved. It is now known that a variety of useful tuning "modes" are available, and in most operating modes the extra coil will ring at a different frequency than the master oscillator.

Primary Switching

Modern transistor or vacuum tube Tesla coils do not use a primary spark gap. Instead, the transistor(s) or vacuum tube(s) provide the switching or amplifying function necessary to generate RF power for the primary circuit. Solid-state Tesla coils use the lowest primary operating voltage, typically between 155 and 800 volts, and drive the primary winding using either a single, half-bridge, or full-bridge arrangement of bipolar transistors, MOSFETs or IGBTs to switch the primary current. Vacuum tube coils typically operate with plate voltages between 1500 and 6000 volts, while most spark gap coils operate with primary voltages of 6,000 to 25,000 volts. The primary winding of a traditional transistor Tesla coil is wound around only the bottom portion of the secondary coil. This configuration illustrates operation of the secondary as a pumped resonator. The primary 'induces' alternating voltage into the bottom-most portion of the secondary, providing regular 'pushes' (similar to providing properly timed pushes to a playground swing). Additional energy is transferred from the primary to the secondary inductance and top-load capacitance during each "push", and secondary output voltage builds (called 'ring-up'). An electronic feedback circuit is usually used to adaptively synchronize the primary oscillator to the growing resonance in the secondary, and this is the only tuning consideration beyond the initial choice of a reasonable top-load.

In a dual resonant solid-state Tesla coil (DRSSTC), the electronic switching of the solid-state Tesla coil is combined with the resonant primary circuit of a spark-gap Tesla coil. The resonant primary circuit is formed by connecting a capacitor in series with the primary winding of the coil, so that the combination forms a series tank circuit with a resonant frequency near that of the secondary circuit. Because of the additional resonant circuit, one manual and one adaptive tuning adjustment are necessary. Also, an interrupter is usually used to reduce the duty cycle of the switching bridge, to improve peak power capabilities; similarly, IGBTs are more popular in this application than bipolar transistors or MOSFETs, due to their superior power handling characteristics. A current-limiting circuit is usually used to limit maximum primary tank current (which must be switched by the IGBT's) to a safe level. Performance of a DRSSTC can be comparable to a medium-power spark-gap Tesla coil, and efficiency (as measured by spark length versus input power) can be significantly greater than a spark-gap Tesla coil operating at the same input power.

Practical Aspects of Design

High Voltage Production

A large Tesla coil of more modern design often operates at very high peak power levels, up to many megawatts (millions of watts). It is therefore adjusted and operated carefully, not only for efficiency and economy, but also for safety. If, due to improper tuning, the maximum voltage point occurs below the terminal, along the secondary coil, a discharge (spark) may break out and damage or destroy the coil wire, supports, or nearby objects.

Tesla Coil Schematics	
 Typical circuit configuration	Here, the spark gap shorts the high frequency across the first transformer that is supplied by alternating current. An inductance, not shown, protects the transformer. This design is favoured when a relatively fragile neon sign transformer is used.
 Alternative circuit configuration	With the capacitor in parallel to the first transformer and the spark gap in series to the Tesla-coil primary, the AC supply transformer must be capable of withstanding high voltages at high frequencies.

Tesla experimented with these, and many other, circuit configurations. The Tesla coil primary winding, spark gap and tank capacitor are connected in series. In each circuit, the AC supply transformer charges the tank capacitor until its voltage is sufficient to break down the spark gap. The gap suddenly fires, allowing the charged tank capacitor to discharge into the primary winding. Once the gap fires, the electrical behavior of either circuit is identical. Experiments have shown that neither circuit offers any marked performance advantage over the other.

However, in the typical circuit, the spark gap's short circuiting action prevents high-frequency oscillations from 'backing up' into the supply transformer. In the alternate circuit, high amplitude high frequency oscillations that appear across the capacitor also are applied to the supply transformer's winding. This can induce corona discharges between turns that weaken and eventually destroy the transformer's insulation. Experienced Tesla coil builders almost exclusively use the top circuit, often augmenting it with low pass filters (resistor and capacitor (RC) networks) between the supply transformer and spark gap to help protect the supply transformer. This is especially important when using transformers with fragile high-voltage windings, such as neon sign transformers (NSTs). Regardless of which configuration is used, the HV transformer must be of a type that self-limits its secondary current by means of internal leakage inductance. A normal (low leakage inductance) high-voltage transformer must use an external limiter (sometimes called a ballast) to limit current. NSTs are designed to have high leakage inductance to limit their short circuit current to a safe level.

Tuning Precautions

The primary coil's resonant frequency is tuned to that of the secondary, by using low-power oscillations, then increasing the power (and retuning if necessary) until the system operates properly at maximum power. While tuning, a small projection (called a "breakout bump") is often added to the top terminal in order to stimulate corona and spark discharges (sometimes called streamers) into the surrounding air. Tuning can then be adjusted so as to achieve the longest streamers at a given power level, corresponding to a frequency match between the primary and secondary coil. Capacitive 'loading' by the streamers tends to lower the resonant frequency of a Tesla coil operating under full power. A toroidal topload is often preferred to other shapes, such as a sphere. A toroid with a major diameter that is much larger than the secondary diameter provides improved shaping of the electrical field at the topload. This provides better protection of the secondary winding (from damaging streamer strikes) than a sphere of similar diameter. And, a toroid permits fairly independent control of topload capacitance versus spark breakout voltage. A toroid's capacitance is mainly a function of its major diameter, while the spark breakout voltage is mainly a function of its minor diameter.

Air Discharges

A small, later-type Tesla coil in operation: The output is giving 43-cm sparks. The diameter of the secondary is 8 cm. The power source is a 10 000 V, 60 Hz current-limited supply.

While generating discharges, electrical energy from the secondary and toroid is transferred to the surrounding air as electrical charge, heat, light, and sound. The process is similar to charging or discharging a capacitor, except that a Tesla coil uses AC instead of DC. The current that arises from shifting charges within a capacitor is called a displacement current. Tesla coil discharges are formed as a result of displacement currents as pulses of electrical charge are rapidly transferred between the high-voltage toroid and nearby regions within the air (called space charge regions). Although the space charge regions around the toroid are invisible, they play a profound role in the appearance and location of Tesla coil discharges.

When the spark gap fires, the charged capacitor discharges into the primary winding, causing the primary circuit to oscillate. The oscillating primary current creates an oscillating magnetic field that couples to the secondary winding, transferring energy into the secondary side of the transformer and causing it to oscillate with the toroid capacitance to ground. Energy transfer occurs over a number of cycles, until most of the energy that was originally in the primary side is transferred to the secondary side. The greater the magnetic coupling between windings, the shorter the time required to complete the energy transfer. As energy builds within the oscillating secondary circuit, the amplitude of the toroid's RF voltage rapidly increases, and the air surrounding the toroid begins to undergo dielectric breakdown, forming a corona discharge.

As the secondary coil's energy (and output voltage) continue to increase, larger pulses of displacement current further ionize and heat the air at the point of initial breakdown. This forms a very electrically conductive "root" of hotter plasma, called a leader, that projects outward from the toroid. The plasma within the leader is considerably hotter than a corona discharge, and is considerably more conductive. In fact, its properties are similar to an electric arc. The leader tapers and branches into thousands of thinner, cooler, hair-like discharges (called streamers). The streamers look like a bluish 'haze' at the ends of the more luminous leaders. The streamers transfer charge between the leaders and toroid to nearby space charge regions. The displacement currents from countless streamers all feed into the leader, helping to keep it hot and electrically conductive.

The primary break rate of sparking Tesla coils is slow compared to the resonant frequency of the resonator-topload assembly. When the switch closes, energy is transferred from the primary LC circuit to the resonator where the voltage rings up over a short period of time up culminating in the electrical discharge. In a spark gap Tesla coil, the primary-to-secondary energy transfer process happens repetitively at typical pulsing rates of 50–500 times per second, depending on the frequency of the input line voltage. At these rates, previously-formed leader channels do not get a chance to fully cool down between pulses. So, on successive pulses, newer discharges can build upon the hot pathways left by their predecessors. This causes incremental growth of the leader from one pulse to the next, lengthening the entire discharge on each successive pulse. Repetitive pulsing causes the discharges to grow until the average energy available from the Tesla coil during each pulse balances the average energy being lost in the discharges (mostly as heat). At this point, dynamic equilibrium is reached, and the discharges have reached their maximum length for the Tesla coil's output power level. The unique combination of a rising high-voltage radio frequency envelope and repetitive pulsing seem to be ideally suited to creating long, branching discharges that are considerably longer than would be otherwise expected by output voltage considerations alone. High-voltage, low-energy discharges create filamentary multibranched discharges which are purplish-blue in colour. High-voltage, high-energy discharges create thicker discharges with fewer branches, are pale and luminous, almost white, and are much longer than low-energy discharges, because of increased ionisation. A strong smell of ozone and nitrogen oxides will occur in the area. The important factors for maximum discharge length appear to be voltage, energy, and still air of low to moderate humidity. There are comparatively few scientific studies about the initiation and growth of pulsed lower-frequency RF discharges, so some aspects of Tesla coil air discharges are not as well understood when compared to DC, power-frequency AC, HV impulse, and lightning discharges.

Applications

Tesla coil circuits were used commercially in sparkgap radio transmitters for wireless telegraphy until the 1920s, and in electrotherapy and pseudomedical devices such as violet ray. Today, although small Tesla coils are used as leak detectors in scientific high vacuum systems and igniters in arc welders, their main use is entertainment and educational displays, Tesla coils are built by many high-voltage enthusiasts, research institutions, science museums, and independent experimenters. Although electronic circuit controllers have been developed, Tesla's original spark gap design is less expensive and has proven extremely reliable.

Entertainment

Tesla coils are very popular devices among certain electrical engineers and electronics enthusiasts. Builders of Tesla coils as a hobby are called "coilers". A very large Tesla coil, designed and built by Syd Klinge, is shown every year at the Coachella Valley Music and Arts Festival, in Coachella, Indio, California, USA. People attend "coiling" conventions where they display their home-made Tesla coils and other electrical devices of interest. Austin Richards, a physicist in California, created a metal Faraday Suit in 1997 that protects him from Tesla Coil discharges. In 1998, he named the character in the suit Doctor MegaVolt and has performed all over the world and at Burning Man 9 different years.

Low-power Tesla coils are also sometimes used as a high-voltage source for Kirlian photography.

Tesla coils can also be used to generate sounds, including music, by modulating the system's effective "break rate" (i.e., the rate and duration of high power RF bursts) via MIDI data and a control unit. The actual MIDI data is interpreted by a microcontroller which converts the MIDI data into a PWM output which can be sent to the Tesla coil via a fiber optic interface. The YouTube video Super Mario Brothers theme in stereo and harmony on two coils shows a performance on matching solid state coils operating at 41 kHz. The coils were built and operated by designer hobbyists Jeff Larson and Steve Ward. The device has been named the Zeusaphone, after Zeus, Greek god of lightning, and as a play on words referencing the Sousaphone. The idea of playing music on the singing Tesla coils flies around the world and a few followers continue the work of initiators. An extensive outdoor musical concert has demonstrated using Tesla coils during the Engineering Open House (EOH) at the University of Illinois at Urbana-Champaign. The Icelandic artist Björk used a Tesla coil in her song "Thunderbolt" as the main instrument in the song. The musical group ArcAttack uses modulated Tesla coils and a man in a chain-link suit to play music.

The world's largest currently existing two-coil Tesla coil is a 130,000-watt unit, part of a 38-foot-tall (12 m) sculpture titled *Electrum* owned by Alan Gibbs and currently resides in a private sculpture park at Kakanui Point near Auckland, New Zealand. The most powerful conical Tesla coil (1.5 million volts) was installed in 2002 at the Mid-America Science Museum in Hot Springs, Arkansas. This is a replica of the Griffith Observatory conical coil installed in 1936.

Vacuum System Leak Detectors

Scientists working with high vacuum systems test for the presence of tiny pin holes in the apparatus (especially a newly blown piece of glassware) using high-voltage discharges produced by a small handheld Tesla coil. When the system is evacuated the high voltage electrode of the coil is played over the outside of the apparatus. The discharge travels through any pin hole immediately below it, producing a corona discharge inside the evacuated space which illuminates the hole, indicating points that need to be annealed or reblown before they can be used in an experiment.

Hazards

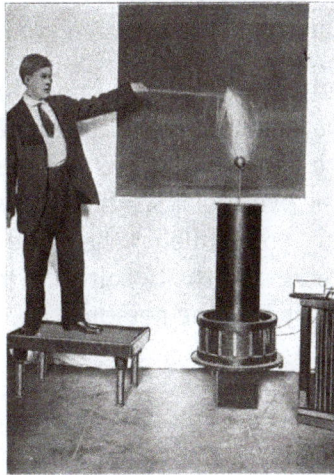

Student conducting Tesla coil streamers through his body, 1909

The 'Skin Effect'

The dangers of contact with high-frequency electric current are sometimes perceived as being less than at lower frequencies, because the subject usually does not feel pain or a 'shock'. This is often erroneously attributed to skin effect, a phenomenon that tends to inhibit alternating current from flowing inside conducting media. It was thought that in the body, Tesla currents travelled close to the skin surface, making them safer than lower-frequency electric currents.

Although skin effect limits Tesla currents to the outer fraction of an inch in metal conductors, the 'skin depth' of human flesh is deeper than that of a metallic conductor due to higher resistivity and lower permittivity. Calculations of skin depth of body tissues at the frequency of Tesla coils show that it can be greater than the thickness of the body. Thus there seems to be nothing to prevent high-frequency Tesla currents from passing through deeper portions of a subject's body, such as vital organs and blood vessels, which may be better conducting. The reason for the lack of pain is that a human being's nervous system does not sense the flow of potentially dangerous electric currents above 15–20 kHz; essentially, for nerves to be activated, a significant number of ions must cross their membranes before the current (and hence voltage) reverses. Since the body no longer provides a warning 'shock', novices may touch the output streamers of small Tesla coils without feeling painful shocks. However, anecdotal evidence among Tesla coil experimenters indicates temporary tissue damage may still occur and be observed as muscle pain, joint pain, or tingling for hours or even days afterwards. This is believed to be caused by the damaging effects of internal current flow, and is especially common with continuous wave, solid state or vacuum tube Tesla coils operating at relatively low frequencies (tens to hundreds of kHz). It is possible to generate very high frequency currents (tens to hundreds of MHz) that do have a smaller penetration depth in flesh. These are often used for medical and therapeutic purposes such as electrocauterization and diathermy. The designs of early diathermy machines were based on Tesla coils or Oudin coils.

Large Tesla coils and magnifiers can deliver dangerous levels of high-frequency current, and they can also develop significantly higher voltages (often 250,000–500,000 volts, or more). Because of the higher voltages, large systems can deliver higher energy, potentially lethal, repetitive high-voltage capacitor discharges from their top terminals. Doubling the output voltage quadruples the electrostatic energy stored in a given top terminal capacitance. Professionals usually use other means of protection such as a Faraday cage or a metallic mail suit to prevent dangerous currents from entering their bodies.

The most serious dangers associated with Tesla coil operation are associated with the primary circuit. It is capable of delivering a sufficient current at a significant voltage to stop the heart of a careless experimenter. Because these components are not the source of the trademark visual or auditory coil effects, they may easily be overlooked as the chief source of hazard. Should a high-frequency arc strike the exposed primary coil while, at the same time, another arc has also been allowed to strike to a person, the ionized gas of the two arcs forms a circuit that may conduct lethal, low-frequency current from the primary into the person.

Further, great care must be taken when working on the primary section of a coil even when it has been disconnected from its power source for some time. The tank capacitors can remain charged for days with enough energy to deliver a fatal shock. Proper designs always include 'bleeder resistors' to bleed off stored charge from the capacitors. In addition, a safety shorting operation is performed on each capacitor before any internal work is performed.

Testing Transformers

Fig. i A 600 kV, 3.33A Testing Transformer for continuous operation
For the production of higher voltages, a number of identical units are put in cascade to add up their voltages as shown in figure ii.

High Voltage power frequency test transformers are required to produce single phase very high voltages. Their continuous current ratings are very low, usually $\approx$ 1A. Even 1A is a very high current rating. This is because a HV test transformer has to supply only the capacitive charging current to the capacitance formed by the dielectric of the test object. However, since the voltage rating requirements are high, the test transformers are required to be produced with very high insulation level. This increases the size of the test sets tremendously. Hence, single units of test transformers are produced maximum upto 700 kV.

Fig. ii A.C. Testing Cascade, 1.2 MV, 1.25 A, short-time operation 4 hours (at an ambient temperature of 35°C). Transformer tanks made of sheet aluminum [TUR Dresden].

Cascaded Transformers

- Cascading a number of single identical units makes transportation, production and erection simpler.

- The cascading principle is illustrated with the basic scheme shown in Fig. iii below in which it can be seen that output of a stage transformer becomes input for the next stage.

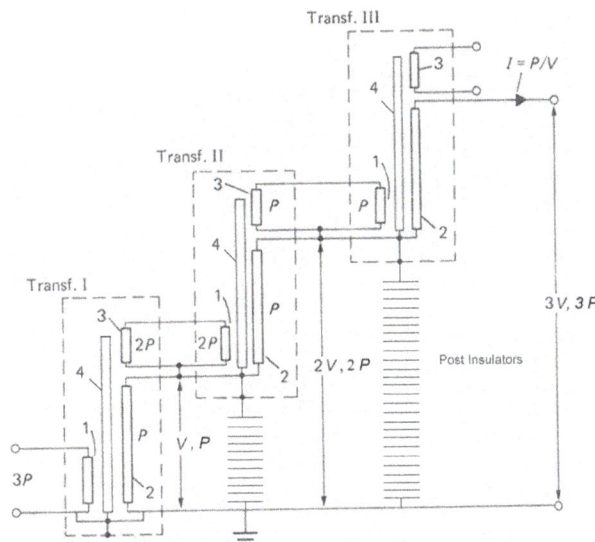

Fig. iii Three Transformers in cascade (1) Primary windings, (2) Secondary, HV, windings, (3) Tertiary/ excitation windings (4) Core

- The HV supply is connected to the primary winding "1" of transformer I, designed for a HV output of V. The other two transformers too are connected in the same fashion.

- The excitation winding "3" of Transformer I supplies the primary voltage for the second transformer unit II; both windings are dimensioned for the same low voltage, and the potential gain is fixed to the same value V.

- The HV or secondary windings "2" of both units are connected in series, so that a voltage of 2 V is produced at the output of 2nd unit. The unit III is added in the same way.

- The tanks or vessels containing the active parts (core and windings) are indicated by dashed lines.

- For a metal tank construction and the HV windings shown in this basic scheme, the core and the tank of each unit would acquire the HV level of the previous unit as indicated . Only the tank of transformer I is earthed.

- The tanks of transformers II and III are at high potentials, namely V and 2 V above earth, and must therefore be suitably insulated, hence raised above the ground on solid post insulators.

- Through HV bushings the leads from the excitation windings "3", as well as the tapings of the HV windings "2", are brought to the next transformer.

For voltages higher than about 600 kV , the cascade of such transformers is a big advantage. The weight and the size of the testing set is sub-divided into single units of smaller size and lower weight. The transportation and erection of the test set in cascade becomes simpler. However, there is a disadvantage that the primary windings of the lower stages are more heavily loaded with higher current in such sets.

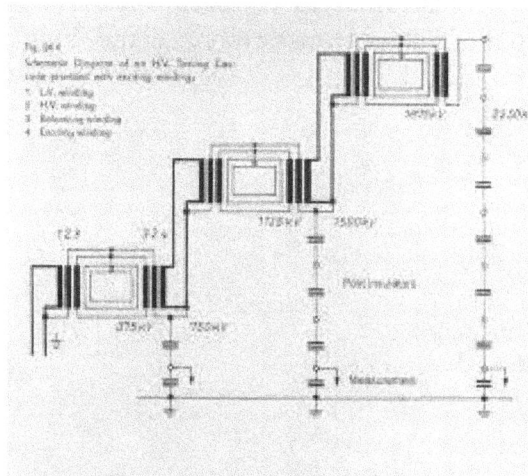

Fig. iv Schematic of an ac test set circuit in cascade

There are several methods of designing the cascade test sets. In Fig. iv schematic diagram of another power frequency test set cascade of 3 x 750 = 2250 kV rating is shown. This circuit has a third winding, known as "Balancing Winding". These windings are designed to acquire the intermediate potentials between two stages. In this circuit, the transformers of the upper stages have their excitation windings arranged over the HV windings of the transformers of the lower potential. Figure v shows a photograph of this test set installed in open air.

Fig. v Photograph of an ac test set of 2250 kV, 2250 kVA installed outdoors

People walking on the ground in this photograph gives an idea of the huge size of the test set.

Testing of HV equipment having high capacitance, for example, long length of HV power cables, power capacitors, GIS etc. may draw excessive capacitive charging current. Necessity for "Tuned series resonant HV power frequency test equipment" arose in particular by the cable manufacturing industry when they required to test long lengths of HV cables drawing large capacitive current on the HV side. Figure vi shows such a circuit. The capacitance C_t represents the capacitance of the test object. A variable reactor is connected on the LV (primary) winding of the test transformer. If the inductance of this reactor is tuned to match the impedence of the capacitive load, the capacitive power can be completely compensated.

The equivalent circuit diagram for this is a low damped series resonant circuit. The high output voltage can be controlled by a variable ac supply, i.e. a voltage regulator transformer (Feed Transformer) if the circuit was tuned before. The Feed Transformer is rated for the nominal current of the inductor and its voltage rating could be very low

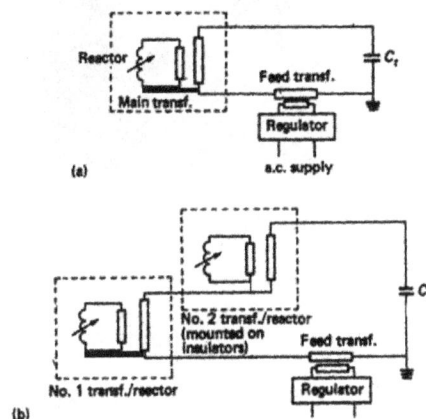

Fig. vi Series resonant circuit for transformer/reactor. (a) Single transformer/reactor.
(b) Two or more units in series.

For simple high capacitive loads the series resonant circuit, shown in Fig. vi and also in Fig vii (a) are suitable. However, for high capacitance and ohmic loads (loads with high real power losses), the parallel resonant circuit shown in Fig vii (b) is more suitable. Both these series and parallel circuits can be made at the same system by changing the connections of the variable reactor 'L'. Fig viii shows a HV variable reactor which is tunned automatically to the desired value of the capacitive load.

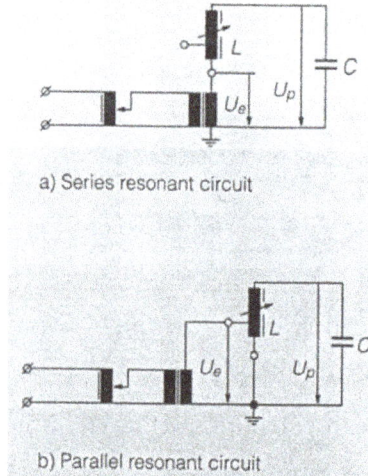

a) Series resonant circuit

b) Parallel resonant circuit

Fig. vii Series and Parallel resonant transformer circuit.

Fig. viii A variable reactor

Advantages of the Series Resonant Circuit

- Dimensions and weight of such test sets are much smaller.

- 100% compensation of capacitive reactive power is possible. Under this condition, the only power drawn from the mains is the active power required.

- The magnitudes of the short circuit currents, in case of insulation failure, are minimised.

- The voltage wave shape is improved by attenuation of harmonic components already in the power supply. A practical figure for the amplification of the fundamental voltage amplitude at resonance is between 20 and 50 times. Higher harmonic voltages are divided in the series circuit to a decreasing proportion across the capacitive load. Good wave shape helps accurate HV measurement and it is very desirable for Schering Bridge measurements.

- The power required from the supply is lower than the kVA in the main test circuit. It represents only about 5% of the main kVA with a unity power factor.

- If a failure of the test object occurs, no heavy power arc will develop, as only the load capacitance will be discharged under this condition. As the voltage collapses, immediately the load capacitance is short-circuited. This has been of great value to the cable industry where a power arc can sometimes lead to the dangerous explosion of the cable termination. It has also proved invaluable for the development work as the weak part of the test object is not completely destroyed. Additionally, as the arc is self-extinguishing due to this voltage collapse, it is possible to delay the tripping of the supply circuit. This results in a recurring flashover condition with low energy level, making it simpler to observe the path of the flashover in air over the surface of the test object.

- By detuning the resonant circuit, breakdown or disruptive discharge can be achieved.

High Direct Voltages

- In HV technology direct voltages are mainly used for pure scientific research work and for testing equipment used in HVDC transmission systems. HVDC test sets are also suitable as mobile test units for testing the equipment at site after installation since these are very light weight.

- High dc voltages are even more extensively used in physics (accelerators, electron microscopy, etc.), electromedical equipment (x-rays), industrial applications (precipitation and filtering of exhaust gases in thermal power stations and cement industry; electrostatic painting and powder coating, etc.), or communications electronics (TV; broadcasting stations). Very high static voltages, produced by electrostatic generators, are used in nuclear physics.

- Therefore, the requirements of voltage shape, voltage level, current rating, short - or long-term stability for every HVDC generating system may differ strongly from each other. With the knowledge of fundamental generating principles, it is possible, however, to select proper circuits for any special application.

- The high dc voltages are generally obtained by means of rectifying circuits applied to ac voltage. Voltage doubler circuits in desired number are then used in cascade for the multiplication of the dc voltage.

References

- Tomar, Anuradha; Gupta, Sunil (July 2012). "Wireless Power Transmission: Applications and Components". International Journal of Engineering Research & Technology. 1 (5). ISSN 2278-0181. Retrieved November 9, 2014

- Mack, James E.; Shoemaker, Thomas (2006). Chapter 15 - Distribution Transformers (PDF) (11th ed.). New York: McGraw-Hill. pp. 15–1 to 15–22. ISBN 0-07-146789-0

- Kubo, T.; Sachs, H.; Nadel, S. (2001). Opportunities for New Appliance and Equipment Efficiency Standards. American Council for an Energy-Efficient Economy. p. 39, fig. 1. Retrieved June 21, 2009

- Allan, D.J. (Jan 1991). "Power Transformers – The Second Century". Power Engineering Journal. 5 (1): 5–14. doi:10.1049/pe:19910004

- AFBI (2011). "9. Contaminants" (PDF). State of the Seas Report. Agri-Food and Biosciences Institute & Northern Ireland Environment Agency. p. 71. ISBN 978-1-907053-20-7

- Bedell, Frederick. "History of A-C Wave Form, Its Determination and Standardization". Transactions of the American Institute of Electrical Engineers. 61 (12): 864. doi:10.1109/T-AIEE.1942.5058456

- Hydroelectric Research and Technical Services Group. "Transformers: Basics, Maintenance, and Diagnostics" (PDF). U.S. Dept. of the Interior, Bureau of Reclamation. p. 12. Retrieved Mar 27, 2012

- Chow, Tai L. (2006). Introduction to Electromagnetic Theory: A Modern Perspective. Sudbury, Mass.: Jones and Bartlett Publishers. p. 171. ISBN 0-7637-3827-1

- Hasegawa, Ryusuke (June 2, 2000). "Present Status of Amorphous Soft Magnetic Alloys". Journal of Magnetism and Magnetic Materials. 215-216: 240–245. doi:10.1016/S0304-8853(00)00126-8

- Figueroa, Elisa; et al. (Jan–Feb 2009). "Low Frequency Heating Field Dry-Out of a 750 MVA 500 kV Auto Transformer" (PDF). Electricity Today. Retrieved Feb 28, 2012

- The Institution of Engineering and Technology (1995). Power system protection. London: Institution of Electrical Engineers. ISBN 0852968361

- Gururaj, B.I. (June 1963). "Natural Frequencies of 3-Phase Transformer Windings". IEEE Transactions on Power Apparatus and Systems. 82 (66): 318–329. doi:10.1109/TPAS.1963.291359

- Manders, Horace (August 1, 1902). "Some phenomena of high frequency currents". Journal of Physical Therapeutics. London: John Bale, Sons, and Danielsson, Ltd. 3 (1): 220–221. Retrieved December 2, 2014

- UNEP Chemicals (1999). Guidelines for the Identification of PCBs and Materials Containing PCBs (PDF). United Nations Environment Programme. p. 2. Retrieved 2007-11-07

- Gottlieb, Irving (1998). Practical Transformer Handbook: for Electronics, Radio and Communications Engineers. Newnes. pp. 103–114. ISBN 0080514561

- Thomson, Elihu (November 3, 1899). "Apparatus for obtaining high frequencies and pressures". The Electrician. London: The Electrician Publishing Co. 44 (2): 40–41. Retrieved May 1, 2015

- Sazonov, Edward; Neuman, Michael R (2014). Wearable Sensors: Fundamentals, Implementation and Applications. Elsevier. pp. 253–255. ISBN 0124186661

- Crookes, William (February 1, 1892). "Some Possibilities of Electricity". The Fortnightly Review. London: Chapman and Hall. 51: 174–176. Retrieved August 19, 2015

- McGinley, Patton H. "Tesla's contributions to electrotherapy" in Childress, David Hatcher, Ed. (2000). The Tesla Papers. Adventures Unlimited Press. pp. 162–167. ISBN 0932813860

- Boteler, D. H.; Pirjola, R. J.; Nevanlinna, H. (1998). "The Effects of Geomagnetic Disturbances On Electrical Systems at the Earth's Surface". Advances in Space Research. 22: 17–27. doi:10.1016/S0273-1177(97)01096-X

Permissions

All chapters in this book are published with permission under the Creative Commons Attribution Share Alike License or equivalent. Every chapter published in this book has been scrutinized by our experts. Their significance has been extensively debated. The topics covered herein carry significant information for a comprehensive understanding. They may even be implemented as practical applications or may be referred to as a beginning point for further studies.

We would like to thank the editorial team for lending their expertise to make the book truly unique. They have played a crucial role in the development of this book. Without their invaluable contributions this book wouldn't have been possible. They have made vital efforts to compile up to date information on the varied aspects of this subject to make this book a valuable addition to the collection of many professionals and students.

This book was conceptualized with the vision of imparting up-to-date and integrated information in this field. To ensure the same, a matchless editorial board was set up. Every individual on the board went through rigorous rounds of assessment to prove their worth. After which they invested a large part of their time researching and compiling the most relevant data for our readers.

The editorial board has been involved in producing this book since its inception. They have spent rigorous hours researching and exploring the diverse topics which have resulted in the successful publishing of this book. They have passed on their knowledge of decades through this book. To expedite this challenging task, the publisher supported the team at every step. A small team of assistant editors was also appointed to further simplify the editing procedure and attain best results for the readers.

Apart from the editorial board, the designing team has also invested a significant amount of their time in understanding the subject and creating the most relevant covers. They scrutinized every image to scout for the most suitable representation of the subject and create an appropriate cover for the book.

The publishing team has been an ardent support to the editorial, designing and production team. Their endless efforts to recruit the best for this project, has resulted in the accomplishment of this book. They are a veteran in the field of academics and their pool of knowledge is as vast as their experience in printing. Their expertise and guidance has proved useful at every step. Their uncompromising quality standards have made this book an exceptional effort. Their encouragement from time to time has been an inspiration for everyone.

The publisher and the editorial board hope that this book will prove to be a valuable piece of knowledge for students, practitioners and scholars across the globe.

Index